Latest Material and Technological Developments for Activewear

Related titles

Lawrence, High Performance Textiles and Their Applications, 978-1-84569-180-6, 2014
McLoughlin, High-Performance Apparel, Materials, Development, and Applications, 978-0-08-100904-8, 2017
Shishoo, Textiles for Sportswear, 978-1-78242-229-7, 2015

The Textile Institute Book Series

Latest Material and Technological Developments for Activewear

Edited by

Joanne Yip

ELSEVIER

WP
WOODHEAD PUBLISHING
An imprint of Elsevier

Woodhead Publishing is an imprint of Elsevier
The Officers' Mess Business Centre, Royston Road, Duxford, CB22 4QH, United Kingdom
50 Hampshire Street, 5th Floor, Cambridge, MA 02139, United States
The Boulevard, Langford Lane, Kidlington, OX5 1GB, United Kingdom

Notices
Knowledge and best practice in this field are constantly changing. As new research and experience broaden our understanding, changes in research methods, professional practices, or medical treatment may become necessary.

Practitioners and researchers must always rely on their own experience and knowledge in evaluating and using any information, methods, compounds, or experiments described herein. In using such information or methods they should be mindful of their own safety and the safety of others, including parties for whom they have a professional responsibility.

To the fullest extent of the law, neither the Publisher nor the authors, contributors, or editors, assume any liability for any injury and/or damage to persons or property as a matter of products liability, negligence or otherwise, or from any use or operation of any methods, products, instructions, or ideas contained in the material herein.

Library of Congress Cataloging-in-Publication Data
A catalog record for this book is available from the Library of Congress

British Library Cataloguing-in-Publication Data
A catalogue record for this book is available from the British Library

ISBN: 978-0-12-819492-8 (print)
ISBN: 978-0-12-822767-1 (online)

Publisher: Matthew Dean
Acquisitions Editor: Brian Guerin
Editorial Project Manager: Fernanda Oliveira
Production Project Manager: Vignesh Tamil
Designer: Victoria Pearson

Typeset by Thomson Digital

Contents

Contributors

Wing-Yu Chan
Institute of Textiles and Clothing, The Hong Kong Polytechnic University, Hung Hom, Hong Kong

Vannesa Chu
Institute of Textiles and Clothing, The Hong Kong Polytechnic University, Hong Kong, China

Elaine Fung
Institute of Textiles and Clothing, The Hong Kong Polytechnic University, Hung Hom, Hong Kong

Olivia Ho-Yi Fung
Institute of Textiles and Clothing, The Hong Kong Polytechnic University, Hung Hom, Hong Kong

Chu-po Ho
Institute of Textiles and Clothing, The Hong Kong Polytechnic University, Hong Kong, China

Garcia Kwok
Institute of Textiles and Clothing, The Hong Kong Polytechnic University, Hung Hom, Hong Kong

Andrew Frank Lando
Founder of ACE Style Institute of Intimate Apparel, The Hong Kong Polytechnic University, Hong Kong; De Montfort University, United Kingdom

Newman Lau
School of Design, The Hong Kong Polytechnic University, Hung Hom, Hong Kong

Nico Liu
Institute of Textiles and Clothing, The Hong Kong Polytechnic University, Hung Hom, Hong Kong

Rong Liu
Institute of Textiles and Clothing, The Hong Kong Polytechnic University, Hung Hom, Hong Kong

Tsz-ling Ng
Institute of Textiles and Clothing, The Hong Kong Polytechnic University, Hung Hom, Hong Kong

Younus Abida
Institute of Textiles and Clothing, The Hong Kong Polytechnic University, Hung Hom, Hong Kong

Lushan Sun
Institute of Textiles and Clothing, The Hong Kong Polytechnic University, Hung Hom, Hong Kong

Kit-lun Yick
Institute of Textiles and Clothing, The Hong Kong Polytechnic University, Hung Hom, Hong Kong

Joanne Yip
Institute of Textiles and Clothing, The Hong Kong Polytechnic University, Hung Hom, Hong Kong

Winnie Yu
Institute of Textiles and Clothing, The Hong Kong Polytechnic University, Hung Hom, Hong Kong

Jun Zhang
School of Design, The Hong Kong Polytechnic University, Hung Hom, Hong Kong

Introduction to activewear

1

Joanne Yip[a], *Andrew Frank Lando*[b,c]
[a]Institute of Textiles and Clothing, The Hong Kong Polytechnic University, Hung Hom, Hong Kong; [b]Founder of ACE Style Institute of Intimate Apparel, The Hong Kong Polytechnic University, Hong Kong; [c]De Montfort University, United Kingdom

1.1 Introduction to activewear

The cynosure of the athletic market was previously focused on men predominantly and marketed function over fashion. In recent years, however, there has been a growing awareness of the benefits of health and wellness, so both men and women are incorporating more fitness into their life. In particular, activewear for women has increased in popularity, and their market share is becoming larger in response to the drive toward a healthy and balanced lifestyle. This means that consumers might seek activewear that would allow a seamless transfer from office to gym and then back again, which has led to the growth of athleisure apparel and the casualization of womenswear. Here, the focus will be primarily on women and activewear. Fashion designers have subsequently collaborated with activewear brands to produce athleisure and intimate apparel to cater to these female consumers. Substantial emphasis has been placed on function and performance in addition to fashion and design in the development of these products. The popularity of athleisure and intimate apparel has also influenced the high-end market, which focuses on superior craftsmanship, new types of fabrics and innovative technologies to meet the needs of consumers who are less price resistant and open to purchasing functional yet aesthetically pleasing products that are an expression of their active lifestyle.

Aside from its increased popularity, activewear is also the only category in apparel that has been experiencing a surge of popularity in recent years. To avoid confusion, activewear will be defined before moving forward. This is followed by a discussion on the main categories and different types of activewear. Then, an overview of activewear in the different markets and the major players in the world is provided. Finally, some of the key factors that influence the purchase intentions of activewear are also reviewed and summarized.

1.2 Definition of activewear

In a search of the existing literature, a broad definition of activewear can be found, which is "some link to being active in the design purpose of the garments" (O'Sullivan, Hanlon, Spaaij, & Westerbeek, 2017). Activewear has been therefore broadly defined as clothing designed to transition from leisure to casual daytime or evening wear. According to Oxford Dictionary, activewear is a noun and defined as "casual, comfortable clothing suitable for sport or exercise" (Simpson, 2017). Nowadays, activewear

Latest Material and Technological Developments for Activewear. http://dx.doi.org/10.1016/B978-0-12-819492-8.00001-6

is designed for specific sporting activities, which can take place outdoors, indoors, in the water, etc., such as rigorous sports, competitive events, physical training sessions, water sports, skiing, and gymnastics. Some of the more popular types of activewear include leggings, tracksuits, golf and tennis shirts, bike shorts, and unitards. These different kinds of apparel are considered to be activewear because they are specifically designed to deliver a pre-defined performance or function to their users, over and above their normal functions (Gupta, 2011).

Another definition of activewear is apparel made for participating in physical activities or sports (Wikipedia, 2019), although it is increasingly worn in daily life for activities such as shopping, going to work, and having a coffee with friends, due to the fusion of fashion and fitness (Fromm, 2016). Activewear is also increasingly becoming a part of streetwear and office wear (Cohen, 2014). It is now considered to be casual and lifestyle wear. Horton, Ferrero-Regis, and Payne (Horton, Ferrero-Regis, & Payne, 2016) described activewear as a form of contemporary fashionable dress that signifies a healthy body and a leisured lifestyle devoted to the improvement of the self. That is, activewear worn casually to work or play implies the casualization of dress and the idolization of a healthy and physically fit body as a fashion aesthetic.

Activewear caters to men, women and children based on market demand. This market is so diversified that it continues to expand and new brands emerge one after another in quick succession.

Activewear is also categorized based on its demanded function, which can be summarized as three different levels of impact performance:

- *Light impact*: refers to activewear for gentle exercise programs, such as yoga and walking, which has the largest market share in sports attire;
- *Moderate impact*: refers to apparel for more aggressive activities, such as cycling, power walking, jogging, dancing, tennis, resistance training and weight lifting; and
- *High impact*: refers to garments for very active and aggressive activities, such as soccer, boxing, kick boxing, lacrosse, field hockey, volleyball, basketball, cheerleading, endurance sports, and cross training.

Nowadays athletes are trained to be competent by focusing on a variety of different exercises, which is known as cross-training. Cross-training gives athletes the ability to adapt to many different types of fitness environments and work different parts of their body. This comprehensive approach to training involves increasing all functional abilities such as stabilization, strength and power. CrossFit is one of the most recent trends in fitness, which is a fitness program created by Greg Glassman in 2000. This fitness regimen is practiced by members in over 13,000 affiliated gyms, and approximately half are located in the US (Glassman, 2002). CrossFit is a core strength and conditioning program. It is not a specialized fitness program but a deliberate attempt to optimize physical competence in each of the 10 fitness domains of cardiovascular/respiratory endurance, stamina, strength, flexibility, power, speed, coordination, agility, balance, and accuracy (Glassman, 2002). CrossFit was developed to enhance competency in all physical tasks. The regimen provides a broad spectrum of activities that create a solid base of knowledge on different sports performances and in choosing the right kind of activewear for the sporting activity. High performance activewear that is found in the high-end market mainly focus on their technically advanced features. Not only do they provide performance and comfort, but also accommodate the lifestyle of their customers.

1.3 Types of activewear

Activewear can be grouped into different sport-specific product categories. Fig. 1.1 shows sketched examples of activewear. It shows some basic design elements in each type of activewear but, it may not include both male and female's styles. A brief description of these activewear examples is also provided below.

Crop tops/crop bras/ workout crop tanks (Fig. 1.1A)—a male/female sleeveless or short-sleeved garment that is cut short to show the midriff (Crop top, 2016). Worn as a top which can be a loose or tight fit depending on the type of sporting activity.

Sports bra (Fig. 1.1B)—the most important piece of activewear for female consumers. Designed to control excessive breast movement and reduce injury to the breasts during vigorous activities. Support and comfort are two important aspects of a sports bra. Sports bras have three major functions: control excessive breast movement, manage heat and moisture, and comfortably support the breasts during sporting activities. Their appearance can be similar to a regular bra or a pull-over bra. Furthermore, the sports bra materials usually resist bacteria (Yu et al., 2006).

Yogawear (Fig. 1.1C)—designed to provide freedom of flexibility, full coverage, compression that slims the body, and performance enhancing features such as moisture-wicking and anti-chafing properties. Yogawear usually consists of a sweat wicking top such as a shirt or tank top for body regulation and a pair of yoga leggings that allow stretch and compression (Madisson, 2019).

Sports pants/leggings (Fig. 1.1D)—a leg covering that usually extends from the ankle to the knee and often made of material such as leather or canvas, which was previously favored for military and work clothes (Legging, 2016). Originally for yoga and specifically pilates, and now fashionable apparel for women or even men. The materials for leggings are usually made of polyester or nylon with high percentage spandex.

Swimwear (Fig. 1.1E)—designed to be worn by those who engage in water-based activities or water sports such as swimming, diving, surfing or activities under the sun. Ranges in exposure from maillot (one piece swimsuit) to two piece bikini. Maillots originated in the 1930s and can have revealing cutouts, such as a backless suit or form-fitting with high cut legs (Calasibetta, 1986). In the 1940s, Louis Rétard invented and named the bikini which went on to become very popular in Europe in the 1950s and the USA in the 1960s (History, 2010).

Tracksuits (Fig. 1.1F)—a two-piece outfit that consists of trousers and a jacket usually with a front zipper worn by athletes during the 20th century. Made of heavy cotton and/or synthetic fibers. The trousers are elasticated at the waist and ankle. The

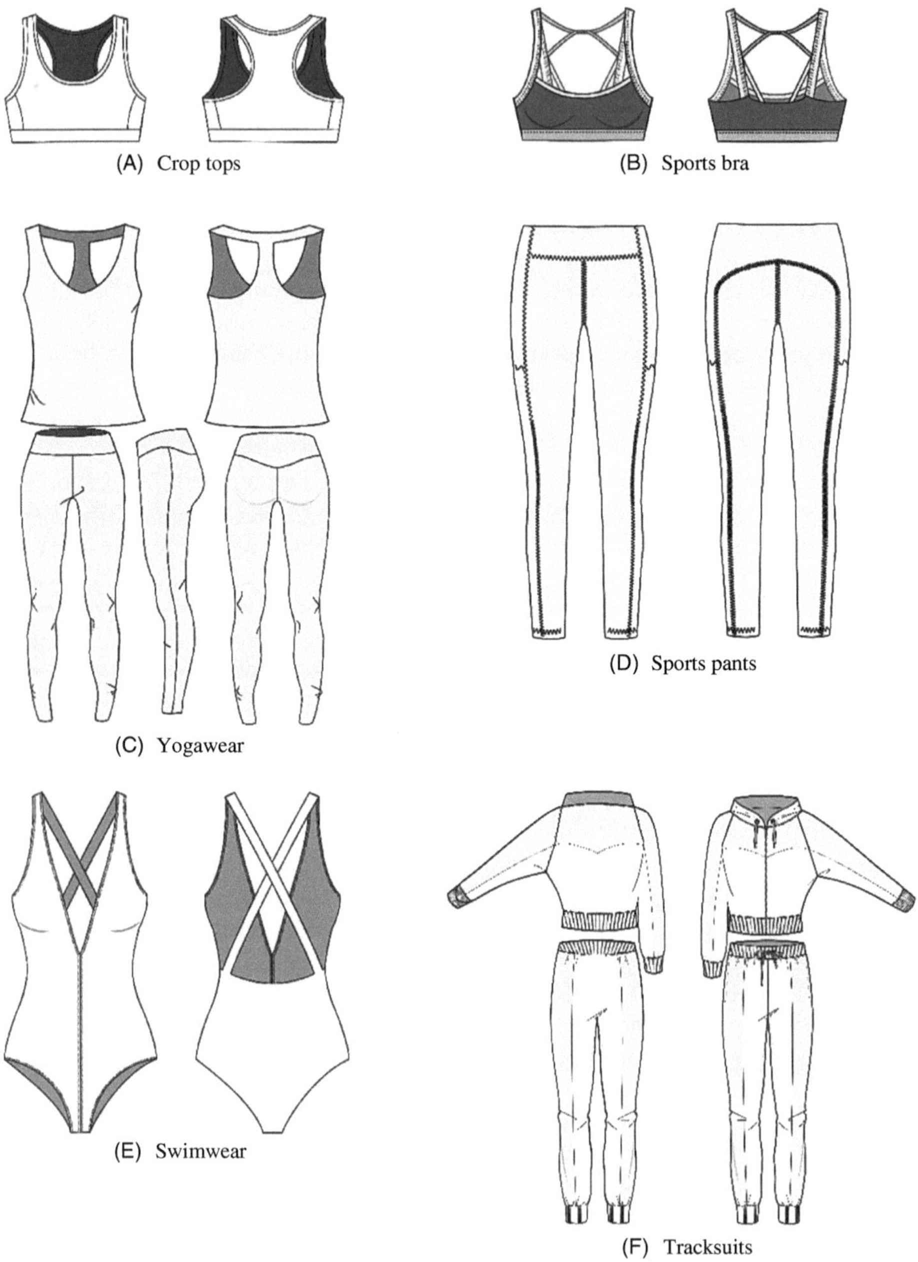

Figure 1.1 Product sketches of different types of activewear.

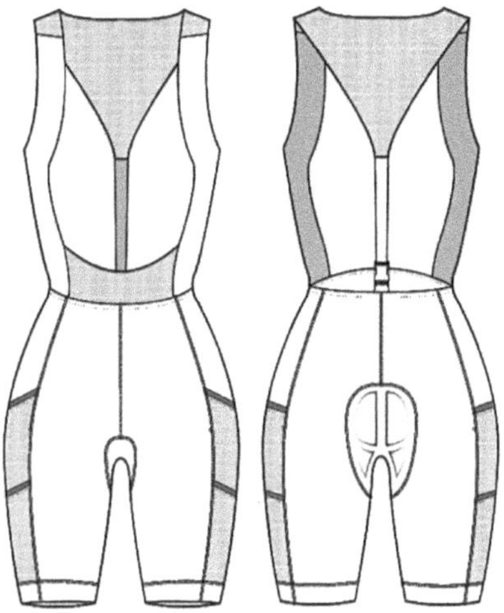

(G) Cycling suits

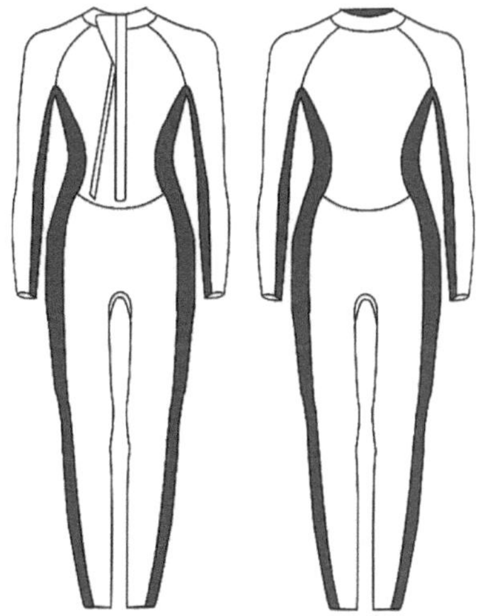

(H) Wetsuits

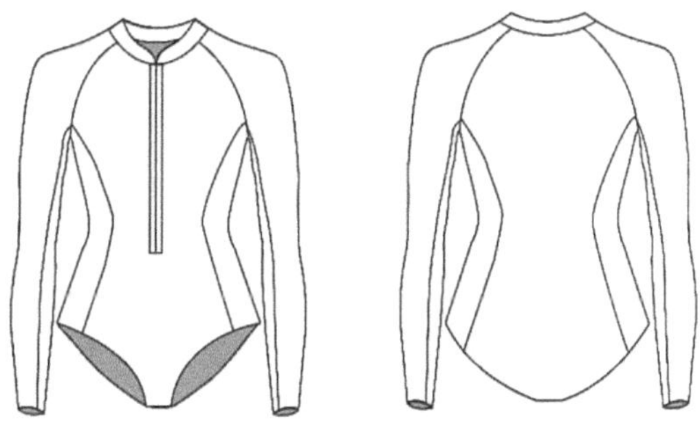

(I) Ski suit- Water sports

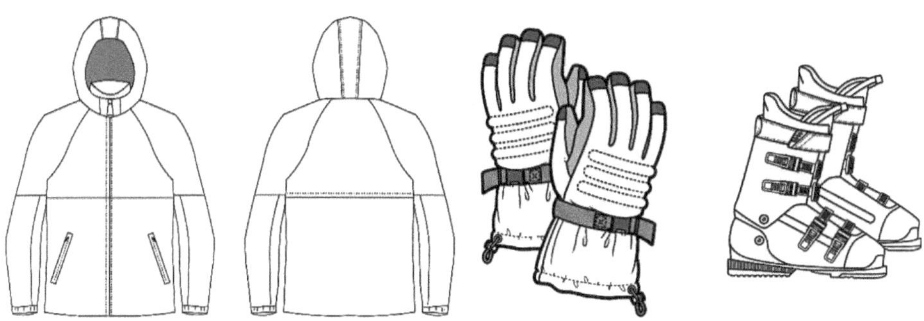

Ski suit with Jacket, Gloves and Boots – Winter sports

Figure 1.1 *(Cont.)*

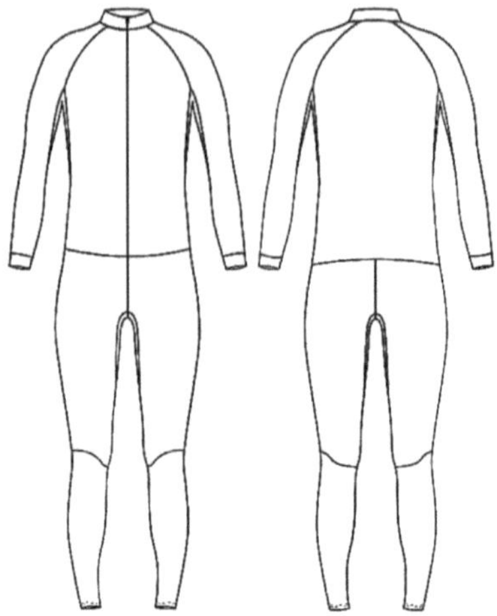

Undershirt of Ski suit – Winter sports

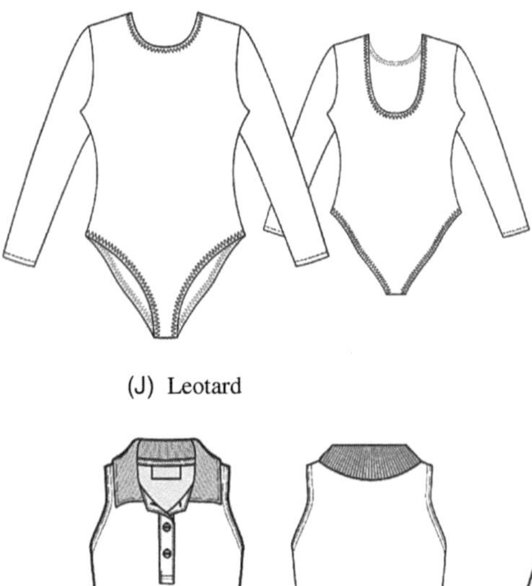

(J) Leotard

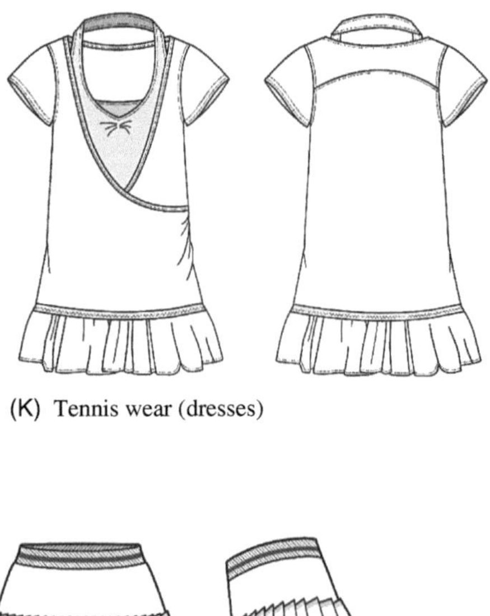

(K) Tennis wear (dresses)

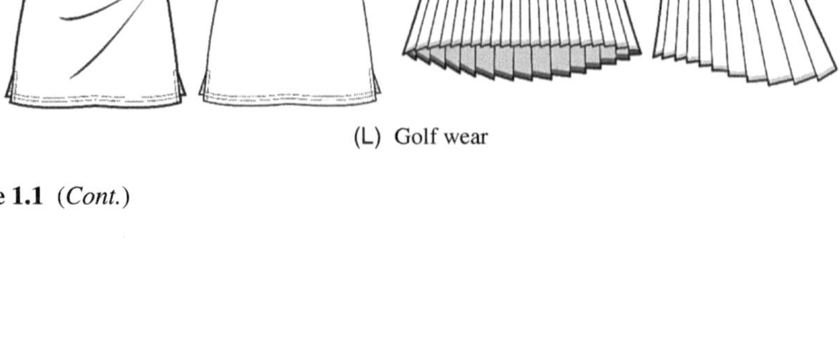

(L) Golf wear

Figure 1.1 *(Cont.)*

top has long sleeves and an elasticated or drawstring waist. In the 1970s when exercise wear became fashionable, the tracksuit was a popular casual attire for women (Heritage, 2008). Worn by many for physical exercise such as jogging.

Cycling suits (Fig. 1.1G)—a type of bike clothing that consists of a jersey and tight fitting shorts that developed into an all-in-one suit for professional racers to provide better air resistance (Ross, 2008).

Wetsuits (Fig. 1.1H)—for scuba divers and fabricated with foam insulation to cover the body and regulate body temperature when diving into deeper water. Covers the body from neck to ankle and made of lightweight non-permeable material, thus giving the wearer more buoyancy and speed in the water. Worn in competitive swimming (Full body swimsuit, 2017).

Skiwear (Fig. 1.1I)—there are two different types of skiwear, one for water skiing which is a regular swimsuit plus a windbreaker to protect the user from the wind while the second one is for snow skiing, when the user will be exposed to extremely cold temperatures on the mountains. The latter can be any type of jumpsuit or two-piece pantsuit which is also known as a snowsuit or snowmobile suit.

- A *snowsuit* is a one-piece jumpsuit or two piece set with an attached hood. The body from the wrists to ankles is covered. Usually worn by infants and toddlers (Calasibetta, 1986).
- A *snowmobile suit* is a monosuit with supplementary features such as snap-close pockets on the chest, reflective stripes on the arms or a self-fabric belt (Calasibetta, 1986).

Skiers will need to be properly dressed in a thick coat or jacket, with gloves and ski boots. Snow jackets are usually made with the fine down of geese or duck for insulation (Calasibetta, 1986). They can also be made with synthetic fibers, namely nylon or polyester fiberfill, which are wind resistant, light and compressible. Also, the undershirt is important, and is known as the second-skin which will need to repel moisture to keep the wearer dry at all times. Ski gear can be very complex and therefore costly as skiing is an expensive winter sport.

Leotards (Fig. 1.1J)—gymnastic wear which is named after a 19th century French acrobat named Jules Léotard, who popularized the garment. The leotard is a one-piece bodysuit that covers the top of the body and the arms to the top of the thighs often with a snap crotch (Calasibetta, 1986). Should fit snugly to allow movement but not so tight that performance is affected. Usually worn by athletes; namely acrobats, dancers or gymnasts.

Tennis wear (Fig. 1.1K)—in 1920, Suzanne Lenglen, a French tennis player introduced the shortened, knee-length, sleeveless tennis dress at a Wimbledon tournament (O'Brien, 1972). Tennis wear is a set of active apparel with different outfit options, from dresses, shorts, sweaters to shoes. Professional tennis players usually require their tennis apparel to have better sweat absorption.

Golf wear (Fig. 1.1L)—similar to tennis wear but the top may be a collared polo shirt. Also includes golf shoes which have cleats to keep the feet on the ground while swinging the golf club and helpful when the ground is wet so that the player does not slip. Golfing gloves provide grip and golf hats keep the sun out of the eyes of the golfer (Calasibetta, 1986).

1.4 Market overview

According to a recently released study called The Future of Apparel by The NPD Group (a leading global information company), activewear currently represents 24% of the total apparel industry sales as of 2018 (NPD Group, 2018). The Les Mils global consumer fitness survey report released in 2019 that was conducted across 22 countries online and received more than 16,000 responses showed that the mean age of fitness club members is 36.4 years old while the mean age of new members is 30.2 years old. Those who are in their twenties (Generation Z) and thirties (the Millennials) will dominate the consumer market as they are estimated to make up two-thirds of the world population, so it is important to take them into consideration (Les Mils Global Consumer Fitness Survey, 2019). They primarily exercise to maintain their shape, look better, be stronger, have fun and feel empowered (Les Mils Global Consumer Fitness Survey, 2019). Therefore, their life style has shifted to a focus on health consciousness and wellness, and pushing the fitness trend.

The growth in the popularity of athleisure wear has changed the views toward workout fashion in that activewear goes beyond exercise and is now a fashion staple. Functional activewear is now trendy and versatile, such as yoga pants, sweatpants or hoodies, which are favored by both female and male fashionistas. Activewear also takes up a large market share as athleisure wear consumers are now the largest market in retail fashion and their future purchasing power in this area is optimistic (NPD Group, 2018).

The Statista Consumer Market Outlook which provides performance indicators of consumer markets globally is now specifically focusing on "Sports and Swimwear" under the apparel category on their website to cater the recent developments in this area in the female apparel industry. In fact, female consumers are becoming more brand conscious since sports brands are creating multidirectional advertising methods for greater exposure and influence so that not only fashion bloggers but also the general public are discussing about these brands. Their channeling methods are usually collaborations with fast-fashion brands or exclusive collections of designer brand, which also create a public relationship with local or even international influencers. As shown in Fig. 1.2, Nike and Adidas are the most purchased brands within the USA, the United Kingdom (UK) and Germany (Statista Research Department, 2019). The results of other market surveys have also shown the impactful influence of athleisure fashion. Consumers in different countries are now indicating that their favorite clothing brand is a sport brand, and indeed, Fig. 1.2 shows that activewear brands are preferred by the majority of consumers.

According to a research firm called Trefis, activewear sales have been growing exponentially in recent years. The estimated total revenue of the global sports apparel market is predicted to increase from 135 billion U.S. dollars in 2012 to 220 billion U.S. dollars in 2024 (Trefis, n.d.).

It is no doubt that the USA is the largest sports apparel and footwear market, with an estimated value of approximately 102.6 U.S. dollars in 2016. Fig. 1.3 shows the largest sports apparel and footwear markets by country in 2016 (Kering, n.d.). Aside from the USA, other important markets are China, Japan, Germany, the UK, France Italy, Brazil, South Korea, and India.

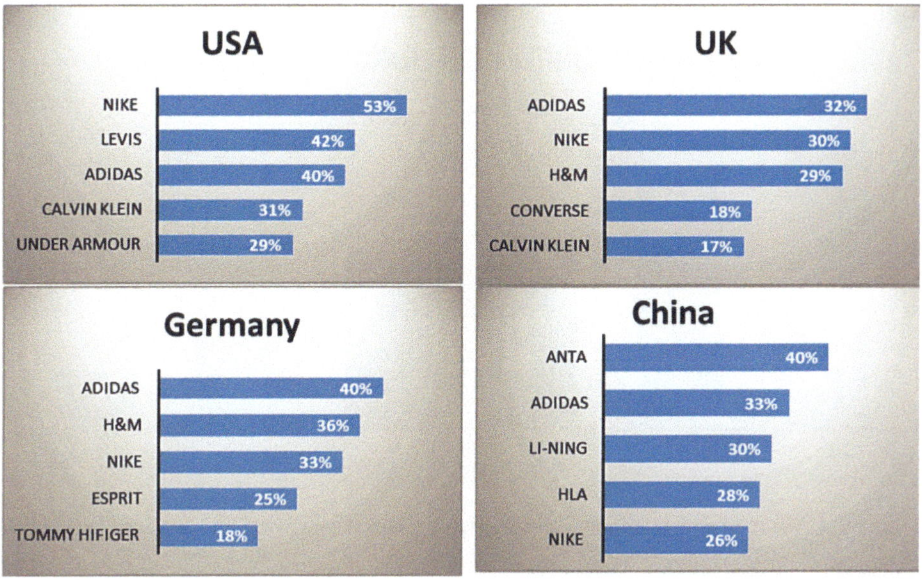

Figure 1.2 Top-5 of clothing, shoe and accessory brands by share of respondents who has purchased them in USA, UK, Germany and China
Source: Statista Research Department (2019).

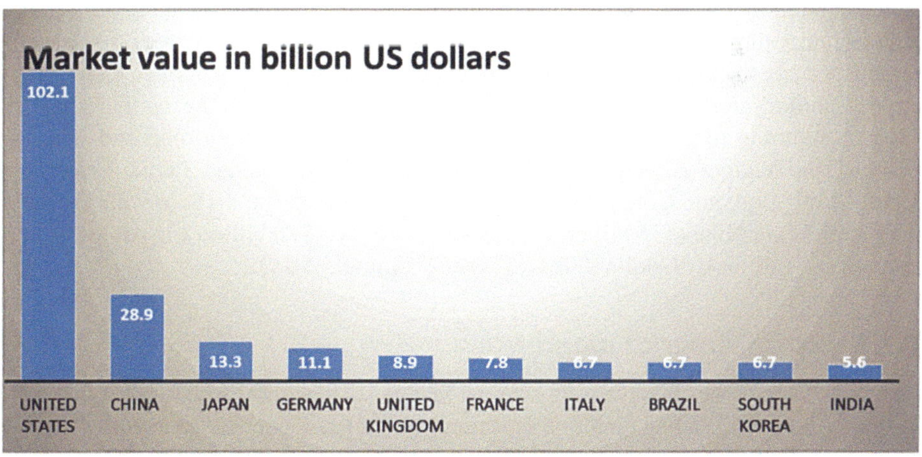

Figure 1.3 Value of sports apparel and footwear market by country: 2016.
Source: Kering (n.d.).

In the sports apparel and footwear market, performance apparel still retains the largest market share, but sports inspired apparel is also growing at a rapid pace. Athleisure is the new trend and the appetite of consumers for athleisure is almost insatiable. Almost every day, a new related business emerges. The market is not only limited to the established sports brands and young designers, but the private labels of retailers

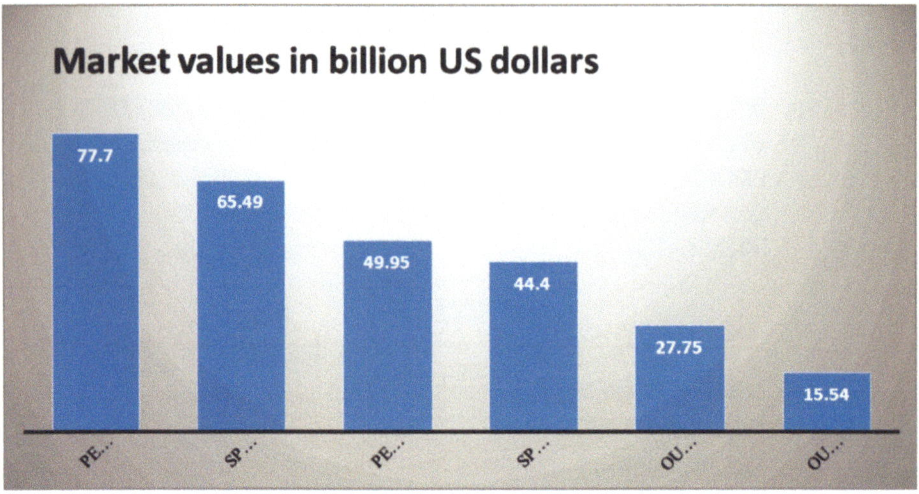

Figure 1.4 Value of sports apparel and footwear market worldwide by product category: 2016.
Source: Kering (n.d.).

are also making their way into this market. Fig. 1.4 shows the value of the sports apparel and footwear market worldwide by product category in 2016 (Kering, n.d.).

The global sports and apparel industry is therefore highly fragmented, with many brands competing, from basic discount brands to high-end fashion names. Even well-established brands have to work hard to maintain their share of the market. Consumers are demanding more versatile wear with more functions, which means that retailers need to continue to provide new sports apparel designs for both men and women. Some of the leading athletic apparel companies are Nike, Adidas, Puma, Under Armour, Lululemon. and Columbia. Fig. 1.5 shows their sales in 2018, in which the top three brands are Nike ($35 billion U.S. dollars), Adidas ($26 billion U.S. dollars). and Puma ($5.2 billion U.S. dollars) (Moss Adams Capital, 2019).

1.4.1 North America market

Because of a number of converging external factors such as shifts in consumer purchasing patterns or the retailing direction of leading labels, the activewear market in the USA is undergoing an unprecedented transformation. According to a census and statistics reports from Cotton Incorporated of the National Cotton Council of America in 2014 which focused on sports apparel, the practicality of athletic wear is becoming a rather insignificant factor for major consumers and the demand for this category of garments has doubled the sales performance of non-active apparel (Cotton Incorporated, 2019).

Apart from this, the key factor for such a phenomenon is due to a growing global awareness in health and wellness. Thus the sales performance and the popularity of athleisure wear reflect this trend in health and wellness, and thus changing the perspective of the public who are now increasingly willing to accept athletic wear

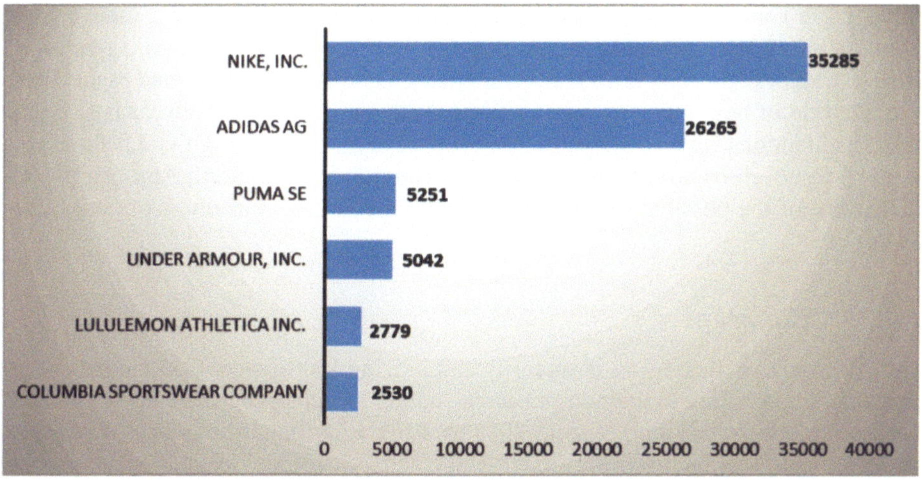

Figure 1.5 Leading athletic apparel, accessories and footwear companies worldwide by sales: 2018 (in millions of U.S. dollars).
Source: Moss Adams Capital (2019).

as a larger part of their fashion wardrobe (Ell, 2019). As consumers increasingly shape their identity around fitness, they prioritize spending time on athletic activities and ensure that they have the proper clothing and equipment to do so, which are reflected by the growing sales of performance apparel and footwear.

In 2019, the revenue of the sports and the swimwear segment in the USA amounted to US$47 billion. According to the revenue breakdown for the different product categories in track and jogging suits, swimwear and functional wear from 2010 to 2023, the average number of pieces of apparel per person amounted to almost four in 2019. In total, this is expected to increase to 1.4 trillion pieces, reaching US$52 billion by 2023 (Lüdemann, 2019).

The activewear market in the USA, which includes both athletic and athletic-inspired apparel and footwear, is expected to increase by US$21 billion by 2023 or a total of over US$138 billion in revenue based on the analytic results from a market research firm called Euromonitor International (Euromonitor International, 2019). During vacation or on weekends, the majority of the general public normally wear some casual wear which is defined as comfortable clothing so that athleisure is also considered to be casual wear. The increase in casualization has been a turning point, in which there is now the trend, that is, typical weekend outfits such as jeans or floral skirts are being traded in for active wear. Hence, an increasing number of brands are creating their own active collections that target audiences who want to be part of the fitness trend by adding yoga pants or other athleisure elements into their closet. This phenomenon has redefined athleisure garments as beyond functional and now fashionable (Euromonitor International, 2019).

Throughout the intensive rivalry between apparel producers and fervent investments in athleisure wear in North America, today the obvious winner is Lululemon

Athletica, a Canadian activewear retailer. Their share price increased almost 80% from 2018 to 2019, and net sales revenue also increased 21% in the third quarter of 2018 to US$747.7 million. This shows the trend of integrating athletic and casual wear into a hybrid of both types of wear in North America. That is, activewear is not only about its functions but also upcycling the product to incorporate high-end fashion elements. Moreover, health and well-being are also priorities in North America, which is also one of the possible reasons that they are the largest consumers of activewear worldwide.

1.4.2 UK market

The UK market is increasing in maturity and moving fast to meet the demands of its consumers. According to previous research studied the changes in women's wear and the active/ leisurewear market from 2013 to 2016. It is found that womenswear has seen only marginal increases in growth while activewear experienced a peak growth of 15.11% for a one-year period from May 11, 2014 to May 10, 2015 (Statista Research Department, 2016). In 2019, the value of the apparel and footwear market in UK was US$77 trillion. Growth has been steady since 2015, and estimated to reach US$81 trillion in 2020, which is about a 16.8% increase in a time span of 5 years (Tiseo, 2019).

New research findings from Rakuten Marketing show that the price ceiling of activewear has increased from £141 per purchase to £187 per transaction on average among 2000 consumers across London and the amount continues to increase across England (Jacks, 2018). Nearly half of the British consumers are likely to have spent this much money on activewear shoes alone—47% admitted that this was the most expensive activewear purchase that they made in 2018 (Jacks, 2018). Another 29% spent more on clothing (Jacks, 2018). The most influential sport when it comes to their activewear purchase decisions is running, in which nearly a fifth of the consumers purchase something related to this sport (19%). Running is also incredibly popular with the younger generation in particular, with a quarter of those surveyed by Rakuten Marketing indicating that it is the most influential sport for their activewear purchase decisions (24%) (Jacks, 2018). This is closely followed by workouts at the gym (16%), which increased to 25% for those between 16 and 29 years old but reduced to 9% for those who are over 60 years old (Jacks, 2018). Almost half or 44% of the respondents indicated that their spending has increased on activewear in the last year in which a quarter of the interviewees are spending most of their earnings on consumption of athleisure items. The largest growth is in the youth market, that is, those who are between 16 and 29 years old (57%) and those who live in the city of London (54%). These are hardly surprising, even if three quarters of the respondents stated that they are most likely to wear activewear for practical reasons, as many people have already taken to wearing activewear outside the gym. Nearly half of the respondents considered activewear as lounge wear and occasionally do wear activewear to do actual workouts. Perhaps, suits or casual smart may not be the only option for the office, as activewear is starting to make its presence there too (Jacks, 2018).

1.4.3 Chinese market

The rate of growth of the Chinese market is supported by government policies that favor active sports, such as participating in world class tournaments or promoting a healthy lifestyle. The government has invested in sports and used entertainment celebrities as spokespersons to endorse sports in order to increase the market support of the younger generation. Sales growth has also increased due to their sponsorship of sporting events too (Rapp, 2017).

The Chinese sportswear market is large and growing at a rapid pace. The retail value of the Chinese market was around US$31 billion in 2017. It is the second largest market after the US market. The Chinese market is growing by 10% each year, and if it maintains a rate of growth of 8% per annum over the next few years, it may reach a total retail value of US$48 billion by 2022 (GMT Research, n.d.).

Fig. 1.6 shows the breakdown of the different brands in the sportswear market in China in 2016 (Fung Business Intelligence Centre, n.d.). The sales of international brands have grown significantly in China. Nike ranked first and Adidas ranked second with a market share of 12.17% and 11.4%, respectively (Fung Business Intelligence Centre, n.d.). During the Singles Day shopping festival in 2018, Alibaba's Alisports branch announced 294 billion RMB (US$43 million) in sales. Women represented up to 43% of those sales, which was up from 33% in 2017 (Ren, 2018).

Local brands like Li Ning has shifted their market share with time, evolving from retailing badminton equipment to becoming a comprehensive sports retailer and a newly emerging fashion label today. One of the most remarkable moves opted by

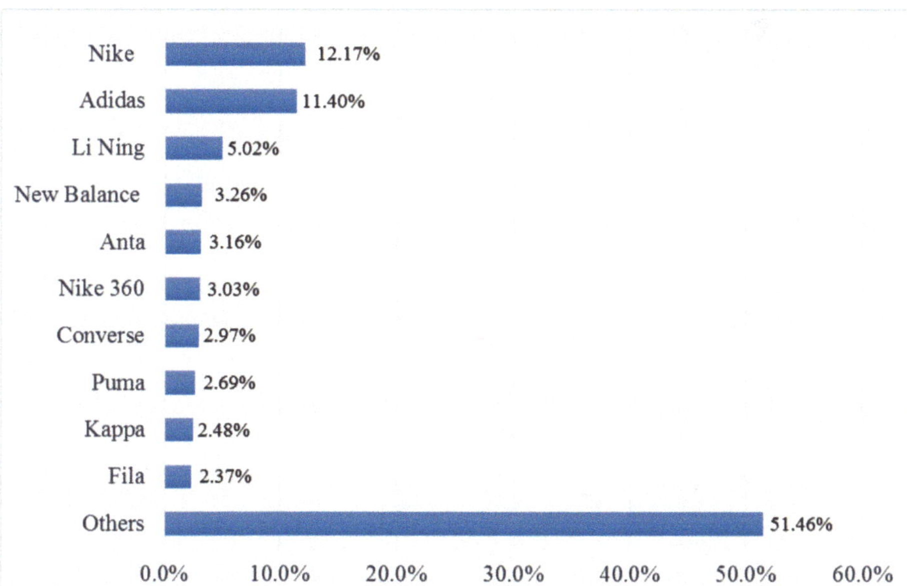

Figure 1.6 Breakdown of sportswear market in China by brand: 2016.
Source: Fung Business Intelligence Centre (n.d.)

Li Ning in 2017 was to redirect the brand to uplifting and sporty yet fashionable garments by signing up key opinion leaders (KOLs) or celebrities in China as their spokespersons to promote exercise in the city (Li Ning Company Limited, 2017). This shows that sports brands are now taking over the fashion world and shaping "athleisure" as a new fashion trend in the 21st century, marked by their presence in the 2018 New York Fashion Week.

Weibo is a very popular social media platform in China and one can gauge the popularity of sportswear brands by looking at their number of followers online. It can be observed that Nike had around 4.28 million followers on Weibo in China. Adidas had 4.03 million followers. Li Ning, 361, and Anta are three local brands but have a good number of followers on Weibo; that is, 627,000, 329,000, and 87,000 followers as of February 2018 (GMT Research, n.d.).

The purchase intentions of Chinese individuals reflect more than just working out. They actually reflect the trend of following an active lifestyle. Adidas provides the impression of strategic and competitive quality at an affordable price similar to their competitors (Xie, 2018). As purchasers are turning to a more healthy lifestyle or immersing themselves into one, the competition between Chinese and international brands has escalated, and local brands aim to transform themselves so that they can compete with their international rivals. Even if the athleisure market in China is still in its infancy, *Particle Fever* has become a leader in this industry. This local brand has taken a very creative approach to marketing and branding. They emphasize that sports bras, leggings, and running shorts are not intended to be just functional, but also be trendy and fashionable. They can compete with their international counterparts because they have a smart branding strategy that positions them as a designer concept store which is precisely responding to the needs of Chinese consumers in modern times (Rapp, 2018).

1.5 Key factors that influence purchase intentions of activewear

Consumers have traditionally considered functionality as the most important purchasing criterion of activewear. However, their needs are now becoming more diversified as there is demand for different products, styles, fabrics, and colors. The activewear industry benefits from this trend by allowing individuals to select their own special type of activewear that meets their design preferences and physical needs. As a result, retailers have the opportunity to expand their market by providing products with the desired functions and design. The purpose of this section is to elaborate some of the key factors that influence the purchase intentions of activewear.

Activewear is apparel worn for an active lifestyle, and since it allows the transition from exercise to casual wear, this type of garment is fast becoming a niche market for both men and women. The term first appeared in the 1916 according to the *Vogue* magazine fashion archives (Condé, 1916), and was linked to the success of aerobics and the workout gear associated with the fitness culture in that era. Today, activewear is a more generic term, in that it is not associated with one sport over another. However, it is still important for marketers to identify the factors that influence the behaviors of their target market (Quester, Pettigrew, Kopanidis, Rao Hill, & Hawkins, 2014). An evidence-based understanding of how activewear is incorporated into people's lives, and their changes in preference of physical activity would benefit the activewear industry.

According to an empirical study by Watts and Chi (Watts & Chi, 2019) on the key factors that influence the purchase intention of American consumers on activewear, attitude and perceived behavioral control showed significantly positive effects on the intention to purchase activewear. Also, past purchase behaviors around activewear and a lifestyle that favors health and wellness help consumers to adopt a positive attitude toward shopping activewear for casual wear purposes. The results of the study by Watts and Chi (Watts & Chi, 2019) indicated that American consumers are more likely to purchase activewear for casual wear purposes if they have a positive attitude to wearing activewear outside of exercise environments. This positive relationship is congruent with the findings from previous apparel purchase intention studies (Wang, 2014; Yan, Ogle, & Hyllegard, 2010; Zheng & Chi, 2015). Therefore, increasingly perceived behavioral control over resources (e.g., money, time, and information) for activewear consumption and self-confidence in wearing activewear outside of exercise environments have led the greater willingness of American consumers to buying activewear.

In contrast, the views and perceptions of other important figures in one's life (e.g., family, friends, and peers) toward activewear consumption do not significantly affect decisions on purchasing activewear. This indifference toward the expectations and opinions of important others suggests that the purchase of activewear is tied to self-interest. Vallerand, Deshaies, Cuerrier, Pelletier, and Mongeau (1992) argued that subjective norms are less significant for explaining purchase intentions because it is a remote concept in this area (Vallerand et al., 1992).

There are differences in preferences within the activewear consumer market. Consequently, psychographic segmentation has been used to identify, understand, and better respond to market sub-segments, instead of simply using demographic segmentation. Many brands use psychographic segmentation to classify products for different needs, including athletic-focused and fashion-focused, from brand to store level. For example, Adidas has two types of stores that offer different types of products; one is Adidas Athletics which is athletic-focused, and Adidas Originals, which is fashion-focused.

In addition, brand personality which focuses on psychographic attributes, has been widely used by activewear brands. That is, consumers not only purchase products for their functions, but also their image (Maehle et al., 2011). Brands have given unique personalities to products to enhance their differentiation strategy and reflect the characteristics of the consumers themselves. Hence, brand communication is designed to have high impacts, yet remains consistent to the theme and message that strongly reflect the brand personality that is suggested to consumers.

Attitudes toward activewear can comprise six aspects, including innovation, fit, durability, design, comfort, and versatility. In the following section, each will be briefly described.

1.5.1 Innovation

Activewear is the culmination of technological improvement, modern fixation on a healthy appearance, and the blurring of fitness fashion and office attire. The driving force behind innovations in activewear is that consumers crave new features, gadgets and increased safety when taking part in sports and are quite prepared to spend large sums of money on high performance activewear. As activewear companies continue to innovate, they have also increased their prices, and this is particularly true of more prominent brands such as Nike, Lululemon, Under Armour, and Adidas.

The history of the Danish textile industry highlights the fact that the significant expansion in the European sportswear industry in the 1950s and 1960s coincided with the development of new materials, that is, synthetic stretch materials that were ideal for sportswear (Bruun & Langkjær, 2016). In 1938, nylon, the first fully synthetic fiber, was developed, and other synthetic fibers and their associated brands such as orlon (acrylic), lycra (polyurethane), terylene (polyester), tactel (adapted form of nylon) and neoprene (synthetic rubber) emerged in the 1950s–1980s. Lycra elastane fiber (spandex), originally invented in the late 1950s, was relaunched in the early 1980s for widespread use by both fashion and sportswear designers to address the new waves of fitness activities, such as aerobics, jazz ballet and bodybuilding.

Today, synthetic technical textiles developed and used for activewear wick away moisture, prevent abrasion and can regulate body temperature, as well as satisfy form requirements that are far more diverse and specialized, particularly since the 1970s. Novelty fabrics also make a brand or line prominent and unique. These high-end fabrics that offer a variety of different functions come from Italy, Japan and Brazil. The examples of innovative fibers and fabrics used in activewear will be described in Chapter 3.

1.5.2 Fit

Fit is one of the most important factors affecting consumer in selecting their activewear. It is required by different body types, shapes, and activities. Garment fit also affects different features of activewear such as breathability and moisture wicking. Moreover, the activewear must be well fitted or secured so that the athletes do not have to adjust it while exercising. The well-designed activewear fit perfectly in the users and remain safe, firm, and flattering throughout the exercise. Therefore, it is important to have the right fit for comfort reason and avoiding injuries.

Since 2016, the average size of American women in general has shifted from a size 14 to 16 which was determined in a study by Christel and Dunn (Christel & Dunn, 2016) that took measurements of a sample of over 5500 women in the United States. The revealed that over the last two decades, the average waist size has increased 2.6 in., from 34.9 to 7.5 in. (Christel & Dunn, 2016). In response to such changes, the major league players of the sportswear market such as Nike and Adidas has also revamped their measurements and expanded their size range to a 22 or XXL. In 2019, Nike used a plus size mannequin in their London flagship store as part of their visual merchandising strategy to actively promote exercise and embrace physical diversity (Dixon, 2019). Therefore, not only these changes redefine how the USA perceives "average" size, but also how women's wear retailers design, market, and sell their clothing to consumers—which could make for shopping experiences with less stress and more options in the future, at least in the USA.

The future of athleisure is moving in more diverse directions, as smaller brands emerge to cater to niche demands and big names shift to accommodate the changes in lifestyle, both in terms of practicality and ideals (Brown, 2018). Catering to consumers across all clothing sizes, in particular, has become crucial as it demonstrates the openness and inclusivity of a brand. For example, Fashercise, a brand based in London, launched a dedicated "Curve" section on its website, which offers brands such as Day Won, Rainbeau, and Ivy Park activewear brands in a wide range of sizes. Another example is ASOS 4505, the fitness line of ASOS itself that ranges from petite to a size 30 under Curve on the Fashercise website. The Good American line offered by influence Khloé Kardashian is a prime example of fashion inclusivity. Good American offers all sizes for every one of its styles, whether it is denim or activewear. They also avoid negative terms that label women as a certain size, such as plus or petite. Finally, consumer can select among three different fit models to see how a certain style would appear on a certain body type on the Good American website.

1.5.3 Durability

Unlike fast fashion in the apparel industry, those who buy activewear often seek quality and are more concerned about the environmental impacts. Fast fashion has been considered as the common and significant contributor to fashion waste due to rapid retailing. Therefore, high-end fabrics are important in activewear, as they ensure a quality product that is durable and sustainable. Research by Yee and Sidek (Yee & Sidek, 2009) concluded that there is a positive and significant relationship between the factors of brand loyalty (brand name, product quality, price, style, promotion, service quality, and store environment) and sportswear brand loyalty.

In contrast, the type of fibrous material, modifications of the yarn or fabric structure or advanced finishing technology such as ultrasonic welding or lamination can also enhance the physical properties of apparel so that their shelf life can be extended. Even if the price of athleisure apparel is high due to those enhancements, they are still worthwhile and highly stimulate the purchase intentions of consumers.

1.5.4 Design

Today, activewear is not only about function or typical features around fit but also appearance itself. A study done by the NDP group noted that the sales of sneakers inspired by running and casual athletic wear have increased by 39% and 24%, respectively (Cheng, 2018). Therefore, it can be concluded that the demand for aesthetics and affiliation has increased at the same pace as activewear itself.

The USA is the leading market segment in activewear retailing with steady growth in sports apparel and footwear. It is estimated that by 2020, sales in activewear apparel and footwear will increase by one third more. This could be associated with the changes in attitude of everyday citizens who are now more actively participating in athletic exercise. Furthermore, celebrities and sports label collaborations have become the trend in street styles or ready-to-go items today, and driven by game changers such as the Yeezy 350 Boost series of sneakers, a collaboration between Kanye West (an American rapper) and Adidas since 2013 (Furness, 2019). There is also the Off-White series of sneakers, a collaboration between Virgil Abloh (an American designer) and Nike since 2017 (Sneaker News Inc, n.d.), or the trendiest collaboration in the market at this moment- the collaboration between Sacai (a Japanese clothing line founded by Chitose Abe) and Nike for the sacai x Nike LDWaffle sneakers in the fall of 2019 (Sneaker News Inc, 2019).

Apart from trainers, the activewear apparel market is the latest target of many designers and seemingly good business. The youngest billionaire who has made her fortune recently in this area is influencer and daughter of Caitlyn (né Bruce) Jenner, Kylie Jenner, who has launched a swimwear series with her sister Kendall in 2016. Both of them are new players in the activewear market. Moreover, sports labels are now seen to be frequently working with celebrities or designers for workout garments. For instance, supermodel Gigi Hadid collaborated with Reebok in 2018 for their campaign (Goldfine, 2018) and American fashion designer Alexander Wang collaborated with Adidas in 2019 (Bobila, 2019). The reaction from the market has been favorable, and so it can be concluded that there is indeed a celebrity effect on the purchase intentions of consumers.

1.5.5 Comfort

Cotton Incorporated in the USA conducts a variety of studies and surveys to inform practitioners and academics on issues related to cotton worldwide. They carried out a survey in 2014 and found that price is the third most important consideration for purchasing clothing in general but drops to sixth for activewear (Cotton Incorporated, 2019). Therefore, less importance placed on price for activewear means that retailers and brands are inclined to put a higher price on such items and benefit from a larger profit margin. However, their ability to continue to charge higher prices depends on whether they can meet the demands of consumers for comfort in activewear.

The survey by Cotton Incorporated indicated close to half of the respondents (42%) rank comfort as the most important factor for their activewear, then fit (16%), breathability (15%), and finally style (14%) (Cotton Incorporated, 2019). Almost three-quarters of the respondents indicated that they wear activewear for any reason other than exercise is because the garments are comfortable. Most stated cotton as the most comfortable (53%) in comparison to synthetic fabrics such as nylon (11%) and polyester (10%), mainly because such garments are not comfortable to wear (Cotton Incorporated, 2019).

1.5.6 Versatility

In the early days, the main purpose of activewear was to protect wearers from their surroundings. With time and technological advancements, the requirements for activewear are not only just for protection but also different functions. For example, resistance to bacteria and odors are some of the common demands since the spaces between fibers provide a warm and humid environment for the growth of microorganisms which causes odor and bacterial growth after the wearer perspires. In addition, protection against wind, rain, and snow in extreme weather conditions is also necessary for activewear like during skiing, snowboarding, and sledding. As for activewear used for outdoor activities such as hiking, fishing and hunting, some other factor namely UV resistance, and soil- and water-repellence are also important to protect wearers from their surroundings. It is therefore recommended that information on the characteristics and properties of products be made readily available to consumers to help them make informed decisions during activewear purchase.

1.6 Conclusion

Activewear is gaining popularity and also only the apparel category that has been experiencing growth in recent years. In this chapter, a definition of activewear is first provided, followed by an overview of activewear in the different markets and the major players in the world. Finally, some of the key factors that influence the purchase intentions of activewear are also reviewed and summarized. It is noticed that consumers have traditionally considered functionality as the most important purchasing criterion of activewear. However, their needs are now becoming more diversified. In many years, innovation, fit, durability, design, comfort, and versatility are important factors that influence purchase intentions of activewear. But recently, some brands have taken a very creative approach to marketing and branding. They emphasize that activewear is not intended to be just functional, but also trendy and fashionable. Many sports brands launch limited collection with celebrities or fashion designers help to boost purchase intention of customers. Organizing some campaigns like Marathon or boot camp or providing specific training through social media also successfully increase the sales of activewear.

In conclusion, the purposes of this book are to introduce new technologies related to design development, material choices, and the manufacture of activewear to the readers, to provide information for designers to design functional, comfortable, and fashionable activewear, to meet the requirements and standards of the apparel and fashion industry, and the most important to educate the next generation on the requirements for activewear along with the modern elements that are incorporated into activewear.

References

Bobila, M. (2019). According to Alexander Wang's Latest Collection with Adidas Originals, Clubbing is a Sport. Available from: https://fashionista.com/2019/04/alexander-wang-adidas-originals-collection-season-five-drop-one, Accessed 7 July 2019.

Brown, H. (2018). Activewear evolution: heading for a diverse, conscious future. Drapers. Available from: https://www.drapersonline.com/product-and-trade-shows/activewear-evolution-heading-for-a-diverse-conscious-future/7031105.article, Accessed 7 July 2019.

Bruun, M. B., & Langkjær, M. A. (2016.). Sportswear: Between Fashion, Innovation and Sustainability. *Fashion Practice*, *8*(2), 181–188.

Calasibetta, C. M. (1986). *Essential terms of fashion: A collection of definitions*. New York: Fairchild Publications.

Cheng, A. (2018). The Athleisure Trend Isn't Taking A Rest. Forbes.Available from: < https://www.forbes.com/sites/andriacheng/2018/02/09/here-are-some-more-signs-athleisure-trend-still-has-legs/#6becabd633cb, Accessed 7 July 2019.

Christel, D. A., & Dunn, S. C. (2016). Average American women's clothing size: comparing National Health and Nutritional Examination Surveys (1988–2010) to ASTM International Misses & Women's Plus Size Clothing. *International Journal of Fashion Design, Technology and Education*, *10*(2), 129–136.

Cohen, M. (2014). "Dressing for Success". *The NPD Group Blog*. Available from: https://www.npdgroupblog.com/dressing-for-success/#.VTSf1qLPuO0, Accessed 30 April 2019.

Condé, N. (1916). Seen in the Shops: The Shops, though in the Midst of a Riot of Frills and Ruffles, Take Practical Thought for the Midsummer Sportswoman. Vogue, 48(2). Available from: https://archive.vogue.com/article/19160715074, Accessed 7 July 2019.

Cotton Incorporated (2019). WINNING IN THE U.S. ACTIVEWEAR MARKET.Lifestyle Monitor. Available from:< https://lifestylemonitor.cottoninc.com/winning-in-the-u-s-activewear-market, Accessed 7 July 2019.

Crop top. (2016). In Editors of the American Heritage Dictionaries, (Ed.), *The American Heritage (R) dictionary of the English language*,.6th ed. Boston, MA: Houghton Mifflin. Available from: https://search-credoreference-com.ezproxy.lb.polyu.edu.hk/content/entry/hmdictenglang/crop_top/0, Accessed 7 June, 2019.

Dixon, E. (2019). Nike introduces plus-size mannequins to London store. CNN Styl, Available from: https://edition.cnn.com/style/article/london-nike-mannequins-scli-intl/index.html, Accessed 7 July 2019.

Ell, K. (2019). Lululemon Is Taking on the Whole World.WWD. Available from < https://wwd.com/business-news/financial/lululemons-five-year-expansion-plan-1203116435/, Accessed 20 June2019.

Euromonitor International. (2019). Country Report: Sportswear in the US. Available from: https://www.euromonitor.com/sportswear-in-the-us/report,Accessed 20 June 2019.

Fromm, J. (2016). The Lululemon lifestyle: Millennials seek more than just comfort from athleisure wear. *The Forbes*. Available from: https://www.forbes.com/sites/jeff-fromm/2016/07/06/the-lululemon-lifestyle-millennials-seek-more-than-just-comfort-from-athleisure-wear/#2fff1f4914d9, Accessed 30 April 2019.

Full Body Swimsuit (2017). *Macquarie Dictionary*,7th ed. South Yarra, Australia: Macquarie Dictionary Publishers.

Fung Business Intelligence Centre (n.d.) Breakdown of the sportswear market in China in 2016, by brand. Available from: https://www-statista-com.ezproxy.lb.polyu.edu.hk/statistics/432292/leading-sportswear-brands-in-china/, Accessed 7 July 2019.

Furness, J. (2019). THE HISTORY OF THE YEEZY BOOST 350 WITH STADIUM GOODS. Farfetch. Available from: https://www.farfetch.com/style-guide/how-to/the-history-of-the-yeezy-boost-350/, Accessed 7 July 2019.

Glassman, G. (2002). The CrossFit Level 1 Training Guide. Available from: <http://library.crossfit.com/free/pdf/CFJ_English_Level1_TrainingGuide.pdf, Accessed 7 July 2019.

GMT Research. (n.d.) Number of Weibo followers for major sportswear brands in China as of February 2018 (in 1,000s). In *Statista - The Statistics Portal*. Available from https://www-statista-com.ezproxy.lb.polyu.edu.hk/statistics/915702/china-number-of-weibo-followers-for-major-sportswear-brands/, Accessed 7 July 2019.

GMT Research. (n.d.) Retail sales value of Chinese sportswear from 2007 to 2022 (in billion U.S. dollar). Available from: https://www-statista-com.ezproxy.lb.polyu.edu.hk/statistics/915626/china-retail-sales-value-of-chinese-sportswear/, Accessed 7 July 2019.

Goldfine, J. (2018). Gigi Hadid's Reebok Collection Is Called 'Future Nostalgia'Available from: http://www.papermag.com/gigi-hadid-rebook-2619754178.html, Accessed 7 July 2019.

Gupta, D. (2011). Functional clothing – Definition and classification. *Indian Journal of Fibre & Textile Research, 36*(4), 321–326.

Heritage (R) dictionary of the English language, 6th ed. Boston, MA: Houghton Mifflin. Tracksuit.,2008. In G.O. Callan, World of art: The Thames and Hudson dictionary of fashion and fashion designers, 2nd ed. London, UK: Thames & Hudson.

History.com (2010). Bikini introduced. THIS DAY IN HISTORY. Available from: https://www.history.com/this-day-in-history/bikini-introduced, Accessed 7 July 2019.

Horton, K., Ferrero-Regis, T., & Payne, A. (2016). The hard work of leisure: Healthy life, activewear and Lorna Jane. *Annals of Leisure Research, 19*(2), 180–193.

Jacks, A. (2018). Gym beats budgeting with Londoners spending nearly £200 per piece of activewear, *The Rakuten Marketing Blog*. Available from: https://blog.rakutenmarketing.com/en-uk/insights/activewear-insights/,Accessed 7 July 2019.

Kering (n.d.). Sports apparel and footwear market value worldwide in 2016, by country (in billion euros). Available from: https://www-statista-com.ezproxy.lb.polyu.edu.hk/statistics/915617/sports-apparel-and-footwear-market-value-by-country-worldwide/, Accessed 7 July 2019.

Kering (n.d.). Sports apparel and footwear market value worldwide in 2016, by product category (in billion euros). Available from: https://www-statista-com.ezproxy.lb.polyu.edu.hk/statistics/915102/sports-apparel-and-footwear-market-value-by-product-category-worldwide/, Accessed 7 July 2019.

Legging (2016). In Editors of the American Heritage Dictionaries, (Ed.), *The American Heritage (R) dictionary of the English language,*6th ed. Boston, MA: Houghton Mifflin. Available from: < https://search-credoreference-com.ezproxy.lb.polyu.edu.hk/content/entry/hmdictenglang/legging/0, Accessed 7 June 2019.

Les Mils Global Consumer Fitness Survey 2019 (2019). Les Mils. Available from: < https://lmimirror3pvr.azureedge.net/static/media/14993/0523755f-48c9-4e0c-a47d-6ab96dd3ae86/les-mills-2019-global-consumer-fitness-survey-key-insights-public-report.pdf, Accessed 7 July 2019.

Li Ning Company Limited (2017). Li-Ning Brand Annual Report 2017 "Li Ning Brand- Developing LI-NING experience value through multi-level and multidimensional marketing campaigns". Available from: http://ir.lining.com/en/ir/reports.php, Accessed 20 June 2019.

Lüdemann, S. (2019). Apparel Report 2019 – Sports and Swimwear Revenue in million US$. Available from https://www-statista-com.ezproxy.lb.polyu.edu.hk/outlook/90030000/109/sports-swimwear/united-states/, Accessed 7 July 2019.

Madisson, G. (2019) The Perfect Yoga Clothes for Women. DOYOUYOGA. Available from: https://www.doyouyoga.com/the-perfect-yoga-clothes-for-women/, Accessed 7 June 2019.

Maehle, N., Otnes, C., & Supphellen, M. (2011). Consumers' perceptions of the dimensions of brand personality. *Journal of Consumer Behaviour*, Available from: https://doi.org/10.1002/cb.355, Accessed 7 July 2019.

Moss Adams Capital (2019). Leading athletic apparel, accessories and footwear companies worldwide in 2018, by sales (in million U.S. dollars) *. Available from:

https://www-statista-com.ezproxy.lb.polyu.edu.hk/statistics/900271/leading-sportswear-and-performance-wear-companies-by-sales-worldwide/, Accessed 7 July 2019.

NPD Group (2018). The Casualization of American Apparel: Will the Athleisure Trend Stay or Go? [Press Release]. Available from https://www.npd.com/wps/portal/npd/us/news/press-releases/2018/the-casualization-of-american-apparel-will-the-athleisure-trend-stay-or-go/, Accessed 14 June 2019.

O'Sullivan, G. A., Hanlon, C., Spaaij, R., & Westerbeek, H. (2017). Women's activewear trends and drivers: A systematic review. *Journal of Fashion Marketing and Management*, *21*(1), ISSN 1361-2026.

O'Brien, J. A. (1972). Documentation of the clothing styles of American women in tennis, golf, and swimming from 1870 to 1970.

Quester, P., Pettigrew, S., Kopanidis, F., Rao Hill, S., & Hawkins, D. (2014). *Consumer behaviour: Implications for marketing strategy*. Australia: McGraw-Hill Education.

Rapp, J. (2017). How Chinese athleisure fans are driving the luxury fashion market. Available from: https://www.scmp.com/magazines/style/fashion-beauty/article/2107025/why-chinese-athleisure-fans-are-driving-luxury, Accessed 7 July 2019.

Rapp, J. (2018). Athleisure wear sales in China grow, but as a fashion trend it is still in its infancy. Available from: https://www.scmp.com/lifestyle/fashion-beauty/article/2142894/athleisure-wear-sales-china-grow-fashion-trend-it-still-its,Accessed 2 April 2014.

Ren, X. (2018). Nike targets increasingly active women's sports clothing market. Available from: http://www.chinadaily.com.cn/a/201812/05/WS5c073f9da310eff30328f29e.html,Accessed 20 June 2019.

Ross, S. (2008). In *Higher, Further, faster: Is Technology Improving Sport?* (Vol. 5). (1st ed.). John Wiley.

Simpson, J. (2017). *Oxford english dictionary* (3rd ed.). Oxford: Oxford University Press.

Sneaker News Inc. (n.d.) Virgil Abloh, Sneaker News Official Website. Available from: https://sneakernews.com/tag/virgil-abloh, Accessed 7 July 2019.

Sneaker News Inc. (2019). The Next Batch Of sacai x Nike LDWaffles Are Coming On September 5th. Sneaker News Official Website. Available from: https://sneakernews.com/tag/virgil-abloh, Accessed 7 July 2019.

Statista Research Department (2016). Percentage change in the women's wear and active / leisurewear market in the United Kingdom (UK) from 2013 to 2016. Available from https://www-statista-com.ezproxy.lb.polyu.edu.hk/statistics/673361/change-in-womens-wear-and-activewear-united-kingdom-uk/, Accessed 7 July 2019.

Statista Research Department (2019). Apparel Report-2019. Statista Consumer Market Outlook - Market Report. Available from: https://www-statista-com.ezproxy.lb.polyu.edu.hk/study/55501/apparel-report/, Accessed 7 July 2019.

Tiseo, I. (2019). Market value of apparel and footwear in the United Kingdom (UK) from 2013 to 2020 (in million euros)**. Available from https://www-statista-com.ezproxy.lb.polyu.edu.hk/statistics/491114/apparel-and-footwear-united-kingdom-uk-market-value/, Accessed 7 July 2019.

Trefis.com. (n.d.) Total revenue of the global sports apparel market from 2012 to 2024 (in billion U.S. dollars)*. Available from:< https://www-statista-com.ezproxy.lb.polyu.edu.hk/statistics/254489/total-revenue-of-the-global-sports-apparel-market//, Accessed 7 July 2019.

Vallerand, R. J., Deshaies, P., Cuerrier, J. P., Pelletier, L. G., & Mongeau, C. (1992). Ajzen and Fishbein's theory of reasoned action as applied to moral behavior: A confirmatory analysis. *Journal of Personality and Social Psychology: Personality Processes and Individual Differences*, *62*(1), 98–109.

Wang, Y. (2014). Consumers' purchase intentions of shoes: Theory of planned behavior and desired attributes. *International Journal of Marketing Studies*, *6*(4), 50–58.

Watts, L., & Chi, T. (2019). Key factors influencing the purchase intention of activewear: An empirical study of US consumers. *International Journal of Fashion Design, Technology and Education*, *12*(1), 46–55.

Wikipedia.com (2019). Sportswear (activewear). Sportswear (fashion). Available from: https://en.wikipedia.org/wiki/Sportswear_(activewear), Accessed 7 July 2019.

Xie, J. (2018). Overseas sports brands enjoy golden age in China. Available from: http://www.globaltimes.cn/content/1129983.shtml, Accessed 2 April 2014.

Yan, R., Ogle, J. P., & Hyllegard, K. H. (2010). The impact of message appeal and message source on Gen Y consumers' attitudes and purchase intentions toward American apparel. *Journal of Marketing Communications*, *16*(4), 203–224.

Yee, W. F., & Sidek, Y. (2009). Influence of brand loyalty on consumer sportswear. *International Journal of Economics and Management*, 2, 221–236.

Yu, W., Fan, J., Harlock, S., & Ng, S. (2006). Innovation and technology of women's intimate apparel. ProQuest. Available from: https://ebookcentral.proquest.com, Accessed 7 June 2019.

Zheng, Y., & Chi, T. (2015). Factors influencing purchase intention towards environmentally friendly apparel: An empirical study of US consumers. *International Journal of Fashion Design, Technology and Education*, *8*(2), 68–77.

Activewear designs and innovations

2

Nico Liu
Institute of Textiles and Clothing, The Hong Kong Polytechnic University, Hung Hom, Hong Kong

2.1 Introduction

Activewear, which is usually worn for physical exercise or as a sportswear, which is conventionally worn for sporting activities, has often been considered to serve lifestyle needs or functional purposes such as allowing freedom of movement, being light in weight, regulating body temperature, and managing moisture with a variety of different garment shapes and constructions based on different functional materials that address and meet specific needs (Garage Gym Power, n.d.). In terms of functional purposes, activewear allows for a better athletic performance and higher perceived level of comfort. However, apart from understanding the needs of consumers and athletes in activewear and providing innovative solutions to meet their needs, some of the garment brands are also acknowledging the fashion component of physical activity so that they can meet the needs of a changing consumer market. That is because consumers nowadays need their clothing to transition more seamlessly from exercise to their social life or even work life. Therefore, there is a demand for fashionable activewear, which would work not only for exercise purposes but also when going out with friends or returning to the workplace after an afternoon workout. By adding more fashionable elements to activewear, consumers do not need to change from a highly functional product into other clothing and can wear the same for multiple purposes (The Business of Fashion, 2015). An Australian-based lifestyle and sportswear attire brand called Female for Life, which targets female consumers, recently conducted a survey and found that the majority of women are more inclined to exercise if their exercise gear is comfortable to wear and flatters their body (Female for Life, n.d.). Their survey results further support the importance of function and comfort in activewear design.

The sports world has long influenced fashion trends; clothing items that start off as functional accessories for sporting activities have now quickly become staples in daily wear wardrobes (House of Coco, 2018). The reason for the overall popularity in activewear corresponds to a heightened interest in fitness and sports, as well as the casualization of the workplace (The Unique Group, 2016). The popularity and profitability of athletic wear have enticed many fashion brands to venture into this market by launching their own activewear collections. They have begun to incorporate functional sportswear and the related elements into their current garment designs, which has ultimately resulted in a surge of fashionable activewear (Nathalie Martin, 2015). Most would agree that activewear is no longer about following a trend or merely a functional product. Indeed, activewear for both men and women need to balance both

Latest Material and Technological Developments for Activewear. http://dx.doi.org/10.1016/B978-0-12-819492-8.00002-8

function and fashion, so the brands can continue to enjoy a competitive advantage (Statista, 2018).

The brands that have a presence in the activewear market include those that are "functional brands" and "fashion brands." The former include well-known sport brands such as Nike, Adidas, Under Amour, Puma, and Lululemon Athletica (Statista, 2018), and some examples of their products will be elaborated in the following sections of this chapter. The latter include some of the most popular fashion brands in the retail industry (Statista, 2018a), including Zara and H&M, and some of their sportswear collections are presented after that. Finally, some of the innovative sportswear products of lingerie brands like Victoria Sport and Wacoal CW-X will also be discussed.

2.2 Functional brands: their innovative products

2.2.1 Nike

Founded in 1964 as Blue Ribbon Sports and then eventually renamed as Nike, Inc. in 1978, the company is one of the largest multinational corporations in the United States (Alexandria Sage, 2008). It is currently the largest designer, developer, manufacturer, and supplier of athletic garments and shoes (Forbes, 2018). Nike products span the following categories: running, basketball, soccer, men and women training, action sports, sportswear, and golf. Their athletic footwear products and sports apparel mainly focus on specific athletic uses but the majority of their products are worn as casual or leisure wear (NIKE Inc, 2017). Nike emphasizes innovativeness and quality in the development and manufacture of its various products.

For instance, Nike first introduced the Flyknit technology in February 2012 and launched the Nike Flyknit sneakers during the 2012 Olympics in London (Highsnobiety, 2016). After using this knitted material, which consists of fibers with high strength to create a series of different sneakers with breathability, targeted support, and stretchability, Nike then launched its first sports bra based on the Flyknit technology in 2017 called the Nike FE/NOM Flyknit Sport Bra (Fig. 2.1), which they described as the "biggest Flyknit apparel innovation yet" (Dan Howarth, 2017). Flyknit incorporates a knitting technology that weaves nylon spandex yarn to provide

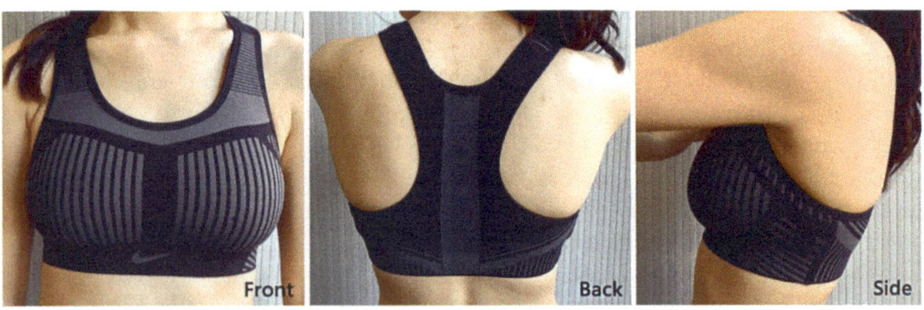

Figure 2.1 Nike FE/NOM Flyknit sports bra.

Figure 2.2 Body map application in NikeLab NIKE A.A.E. 1.0 T-shirt.
Source: NIKE Inc (2017a), and END (2019).

different degrees of support along a piece of single fabric. The Flyknit Sports Bra therefore has the ability to hold each breast by itself, but also compresses the breasts during physical activity (Dan Howarth, 2017). On the other hand, the overall weight of this bra is reduced by about 30 percent due to fewer seams, compared to the other Nike bras (Dan Howarth, 2017; NIKE Inc, 2019). During the product development process, Nike staff evaluated different factors, including thermal effects, perspiration, cooling, and areas of movement of the body with the use of motion capture and digital body scanning, in order for the most affected parts of the body to receive adequate support, and allow breathability and cooling (Dan Howarth, 2017; NIKE Inc, 2019; Vogue, 2017).

Nike also released the Data-Driven "Advanced Apparel Exploration" (A.A.E.) 1.0 Collection in 2017, which uses a data-driven design to follow the human form and function with an organic appearance for urban wear. The NikeLab NIKE A.A.E. 1.0 T-shirt (Fig. 2.2) applies a computational design to amalgamate different body maps to form a motion-driven pattern, which include cling, sweat, and heat maps (NIKE Inc, 2017a; Alice Morby, 2017). The body maps are based on the reactions of the body to a variety of different modern urban environments, which can have significant fluctuations in temperature and humidity (Robert Lewis, 2017). These transcend the conventional way of constructing a t-shirt. Apparel designers and Nike Flyknit staff have also worked together to merge motion and temperature maps into a single dataset that translates into a visual language (NIKE Inc, 2017a).

2.2.2 Adidas

Another large designer, developer, manufacturer, and supplier of athletic garments and shoes is Adidas AG in Germany. In the 2000s, Adidas was the largest sportswear manufacturer only behind Nike globally and the largest in Europe (Robert Lewis, 2017). Several subbrands fall under the company, including Adidas Sport Performance, Adidas Originals, and Adidas Sport Style. Adidas Sport Performance emphasizes Adidas as a multisport specialist, and focuses on innovation and technology in competitive sports, including basketball and football so that the "impossible" can be achieved (Group of BATGMX, 2013). On the other hand, Adidas Originals and Adidas Sport Style target the sports lifestyle and fashion market (Skemateam7, 2013).

Figure 2.3 Idea of cooling spheres in Adidas Climachill Chill Tee.
Source: Adidas (2019a).

Of particular significance is Climachill, which was launched in 2014. Climachill is a revolutionary active cooling type of apparel. In this apparel line, the combination of high tech fibers and cooling spheres allows heat to be transferred away from the body and produces a cooling feeling so that the wearer can train more intensively and run for a longer period of time. The cooling spheres are placed in areas of the body that have the highest heat production, such as the neck, to reduce the body temperature (Daniel Mitchell, 2004). Also included are microfibers, which serve the purpose of a mesh screen to wick moisture away from the body (Daniel Mitchell, 2004). As a result, the wearer feels comfortable because s/he can stay dry and cool and focus on the physical activity itself. The Climachill series includes a training t-shirt called "Chill Tee," which keeps the wearer feeling cool and dry, as shown in Fig. 2.3.

Adidas countered the Flyknit technology of Nike with adizero Primeknit, which was released in 2012 at the Olympics in London (Knitting Industry, 2012). Primeknit has been widely applied to a variety of different sneaker series, including Ultra Boost and Tubular X Primeknit (Knitting Industry, 2012). Extending its business boundaries, Adidas then proceeded to release its first Primeknit apparel collection called "Primeknit Pack" in January 2016, which included EQT 2/3 F15 and EQT 3/3 F15 sneakers as well as a pair of pants and a jacket (see Fig. 2.4) (Hyperbeast, 2015). The outerwear of the Primeknit Pack boasts a seamless knit material throughout (Highsnobiety, 2016) and the jacket also has hidden zippers, a drawstring hood, and a stand-up collar (Hyperbeast, 2015). The Primeknit technology has also been subsequently applied to other apparel items like tights (Fig. 2.5) (Adidas, 2019b).

2.2.3 Under amour

Another American company that produces performance apparel, sports goods, shoes, and casual wear is called Under Armour, which started with the launch of t-shirts that were light in weight yet also allowed wearers to stay dry and cool in

Figure 2.4 Sketch of a jacket from Adidas "Primeknit Pack" collection.
Source: Highsnobiety (2016) and Sneakersnstuff and Sneakersnstuff Inc (2019).

1996 (Robert, 2005). The t-shirts used moisture-wicking synthetic microfibers, and other garments were no match for its speed in wicking away perspiration (Under Armour, 2019b). The innovative Moisture Transport System pulls perspiration from the body of the wearer (St. James Press, 2004). Other major brands like Reebok and

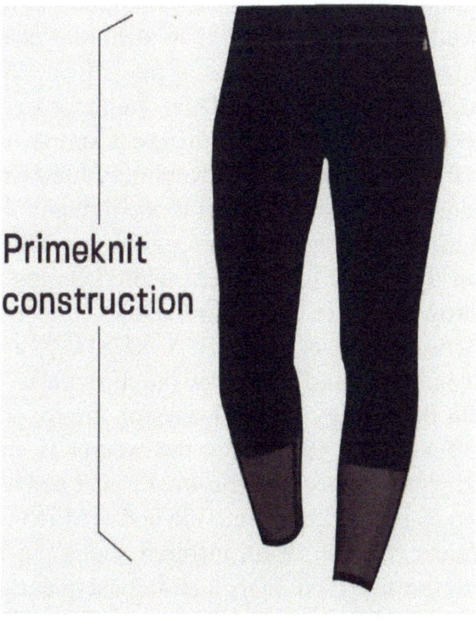

Figure 2.5 Sketch of "Believe This Primeknit LTD Tights".
Source: Adidas (2019b).

Figure 2.6 Sketch of under Armour AllSeasonGear reversible fitted crew shirt.
Source: Fighter X Fashion (n.d.).

Adidas soon came out with their own products that could wick away perspiration (Fast Company, 2005). However, as the pioneer of this technology, Under Armour continues to advance their technology so that wearer performance excels at every level (St. James Press, 2004).

In 1997, Under Armour released its AllSeasonGear line, which is fabricated with radical fabrics that can flex in accordance with changes in temperature and conditions (Under Armour, 2019a). The line has a moisture transport system that can regulate the core temperature of the wearer and allows maximum breathability (Under Armour, 2019c). These properties help to increase the comfort level of wearers when they are exercising outdoors. AllSeasonGear has a unique fabric construction because there is a balance allowed to address a variety of different changing conditions, and it is designed to accommodate temperatures that range from 55°F to 75°F or 13°C to 24°C (Sports Unlimited, n.d.; Riley Jones, 2013). Built for lightweight performance, the AllSeasonGear line targets a variety of different sports, including running and golfing. Fig. 2.6 shows the AllSeasonGear Reversible Fitted Crew Shirt. This reversible shirt allows the wearer to flip the shirt over to the lighter colored side so that when worn against the skin during training, the user feels cooler. The shirt can also be worn with the darker color on the inside to trap heat while keeping dry and feeling warm when the temperature drops (Fighter X Fashion, n.d.).

The latest Under Armour line is called UA RUSH™, a new line of workout gear that was launched in 2019, and uses innovative material that has been infused with minerals to recycle the energy of the body for improved performance (Under Armour, 2019d; Wear Testers, 2019). When the wearer is exercising, the material absorbs the heat emitted from the body of the wearer and converts the heat into infrared energy, which is then re-emitted back into the body (Mirror, 2019). Under Armour claims that this is a "fabric version of an infrared sauna" in that circulation of the body is increased due to the recycled energy and consequently increases the performance, stamina, vigor, and recovery of the wearer; see Fig. 2.7 (Mirror, 2019; Under Armour, 2019e).

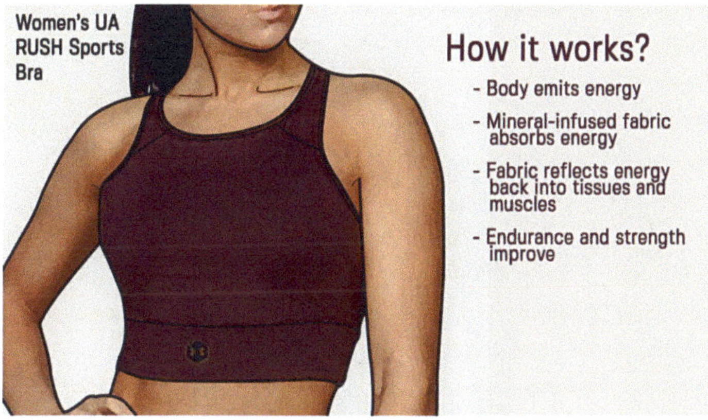

Figure 2.7 Mechanism of mineral infused fabric for Under Armour RUSH garments.
Source: Under Armour (2019e).

2.2.4 Puma

Puma SE, or better known as its brand name Puma, is also a German athletic and casual footwear designer and manufacturer, who has been offering innovative products since 1948 (Puma SE, 2017). Some of the well-known Puma product lines include its Performance apparel and footwear, which are designed for running and football (Puma SE, 2017). Puma's motto is: "Not to create for the sake of creating, but with the purpose to take sport to a different level and to keep moving forward, grounded by science" (Puma SE, 2018a).

Puma launched its Nightcat Powered Jacket in 2014, which was mainly designed for runners who run at night, with the aim to increase their safety in dark environments during training by offering excellent visibility (Puma SE, 2018b). Unlike a normal reflective jacket, PUMA incorporated light-emitting diode (LED) lights into the jacket structure and Bluetooth ability so that the jacket can be used with a mobile phone. The idea is to provide the user with a running partner. The jacket is not only intended to keep the wearer safe, but also provides direction and ability to connect and compete with other runners (Puma SE, 2018b). Imbedded into the back and arms of the jacket is a fiber-optic cable that is flexible and nonirritating. The cable is a removable and rechargeable LED light unit in the pocket of the arm (Men's Running, 2013). Also, there are holes that are laser cut from the back of the jacket to allow more air to circulate in areas that are considered to be heat zones. The preformed sleeves also allow a flexible fit and freedom during running. Fig. 2.8 shows the Puma Nightcat Powered Jacket and its design details (Puma SE, 2018b).

In 2016, Puma launched the EVO TRG jacket, which is a football jacket that offers dynamic thermoregulation to regulate the body temperature of the athlete (Just-style, 2016). The jacket adapts to how the wearer moves and enhances his/her performance with its novel ventilation system that uses a technical stretch mesh (Puma SE, 2018b; Just-style, 2016). The ventilation areas on the jacket open when the wearer moves, which allow the entry of cool air but exit of warm air to regulate the body

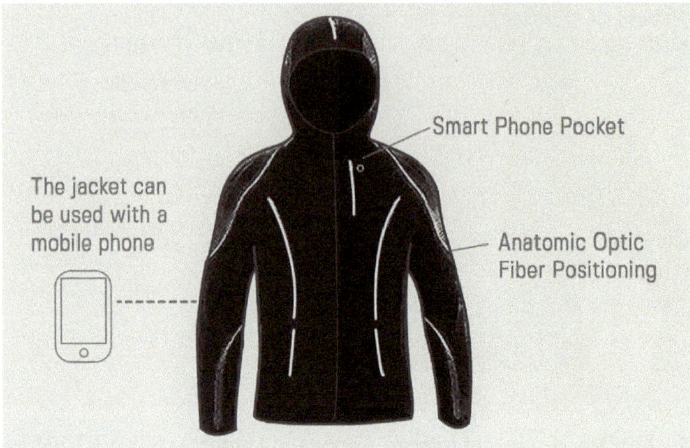

Figure 2.8 Sketch of Puma Nightcat Powered Jacket with design details.
Source: Puma SE (2018b).

temperature; see Fig. 2.9 (Puma SE, 2018b). During the process of developing the jacket, the Puma team identified the prominent heat zones and carefully analyzed how the garment moves and the parts that are stressed during movement (Puma SE, 2018b). The evaluation of the performance of the EVO TRG jacket is not only based on its design process alone but also validated by using thermo-sensing cameras.

2.2.5 Lululemon Athletica

One of the more recent athletic apparel retailers, which has met considerable success, is Lululemon Athletica Inc., a company based in Canada and founded in 1998. While

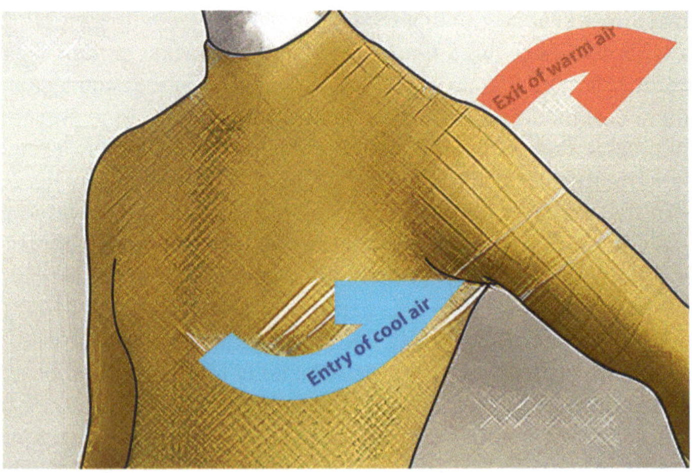

Figure 2.9 Idea of innovative ventilation system on Puma EVO TRG jacket.
Source: Puma SE (2018b).

the company now targets both men and women (Michael Angelo Quiroz, 2018), initially they designed yoga wear for only women (Lululemon Athletica, n.d.). However, after receiving feedback from athletes, users, and the general public, Lululemon has extended their designs to yoga, cycling, etc., for both men and women (Lululemon Athletica, n.d.). They also offer a variety of different types of athletic wear, including performance shirts, shorts, and pants, as well as lifestyle apparel and yoga accessories. Their success mainly lies in the quality and innovativeness of their products (Michael Angelo Quiroz, 2018). Lululemon has a specific research lab called Whitespace, which carries out research and embarks on new developments to advance materials and technologies for new products (Michael Angelo Quiroz, 2018; Chicago Tribune, 2018). As a result, they have created innovative materials such as Luon, Silverescent, Swift, Luxtreme, Nulu fabric, and Everlux fabric (Chicago Tribune, 2018; Michael Angelo Quiroz, 2018).

Luon is the original four-way stretch fabric in the Lululemon family and primarily meant for yogawear (Lululemon Athletica, n.d.; Michael Angelo Quiroz, 2018). Luon has a high proportion of nylon microfibers, which wick away perspiration and provide a soft handfeel (Lululemon Athletica, n.d.). Consumers fell in love with the ability of Luon to recover and stretch infinitely (Lululemon Athletica, n.d.). Aside from Luon, there is Seriously Light Luon, which is a very light version of Luon for hot climates, Light Luon, which can be layered, and Full-On Luon, a tightly knitted version with more support and coverage (Lululemon Athletica, n.d.). Fig. 2.10 shows some examples of Lululemon's yoga pants of Lululemon.

The Silverescent line is another example of a Lululemon fabric technology, which is a material that resists odors and bacteria (Michael Angelo Quiroz, 2018). The silver deposited onto the fabric releases positive ions and resists the bacteria that cause odors that come with perspiration (Lululemon Athletica, n.d.). Silverescent uses the X-STATIC technology, in which 99.9% pure silver is bonded onto the surface of each fiber. The silver will never wash out or cease to work (Lululemon Athletica, n.d.). Silverescent has been applied to different Lululemon products including the short

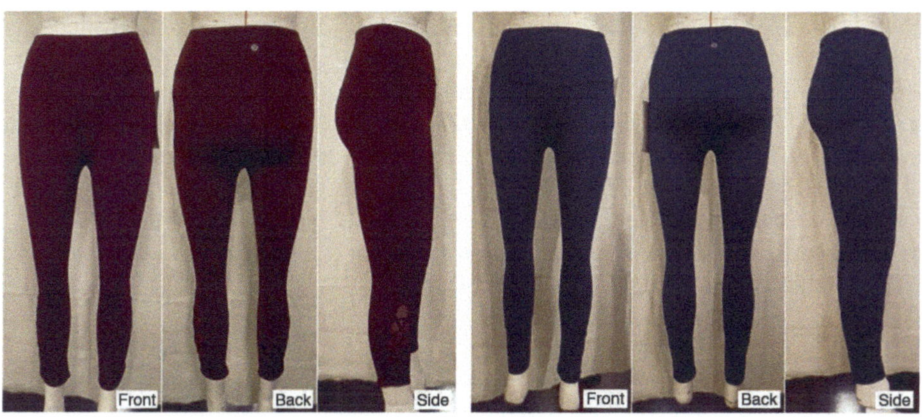

Figure 2.10 Examples of Lululemon's yoga pants.

sleeve crew for both men and women, racerback vest for women, and long sleeve top and polo shirt for men.

Besides, Lululemon has developed the "Sensations" apparel series to meet the needs of their customers for perceived comfort and support. These products include Tight Sensation, Held-in Sensation, Hugged Sensation, Relaxed Sensation, and Naked Sensation (Lululemon Athletica, n.d.). They deliver different performances with the use of different materials and garment constructions (Lululemon Athletica, n.d.), which exert different levels of compression to the body of the wearers during training.

2.3 Functional brands: from function to fashion

Traditionally, functional sportswear brands have been associated with athletes. Due to changes in the market, these brands have incorporated fashion elements into their sportswear in recent years. The reason that leading functional brands are now including fashion elements is attributed to the overall growth of the activewear market, especially in terms of women's business attire. Women want sportswear products that have a wider color range and greater versatility to give them new looks and new experiences. Former Nike President Trevor Edwards said that "In the women's business, there is no performance without style" (The Business of Fashion, 2015).

2.3.1 Nike

In the past, Nike has usually favored athletes over fashion models for its seasonal campaigns. However, the company changed direction and worked with Karlie Kloss, one of the top fashion models in the 2000s, for their Winter 2014 campaign to promote a modern interpretation of sportswear for women (The Business of Fashion, 2015; Cat Meffan, n.d.). As a former ballet dancer and current yoga practitioner, Karlie is associated with both sports and fitness (Fashionista, 2014), but more importantly, she is synonymous with fashion. To further prove its capability in producing fashionable sportswear, Nike presented the "NikeLab x JFS" collection in 2015, which was produced in collaboration with a German designer called Johanna F. Schneider (The Business of Fashion, 2015). The collection featured outfits that can be built, layer by layer, for different stages of a workout (NIKE Inc, 2015). The garments are designed to be fully adaptable as their volume and position can be changed on the body for the different stages of use (NIKE Inc, 2015). In the same year, Nike also hosted a fashion show in New York that celebrated the "Women's Innovation Summit" to unveil its latest endeavors in collaboration with a Brazilian designer, Perdo Lourenco (The Business of Fashion, 2015; So Sasha, 2014).

2.3.2 Adidas

To compete with Nike, Adidas worked with a number of fashion designers including Belgian fashion designer Raf Simons and Greek fashion designer Mary Katrantzou

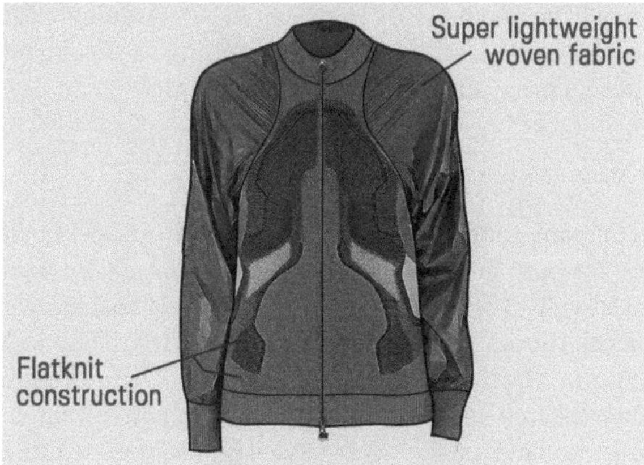

Figure 2.11 Sketch of a jacket in Spring Summer 2019 collection—Adidas by Stella McCartney.
Source: Stella Mccartney (2019).

(The Business of Fashion, 2015). These working partnerships added to the longstanding collaborations that were developed with English fashion designer Stella McCartney for Adidas by Stella McCartney and Japanese fashion designer Yohji Yamamoto for Y-3. The partnership with Mary Katrantzou for Adidas Original consisted of the Originals by Mary Katrantzou collection, which was an innovative apparel and shoe collection delivered in 2014, and incorporated her iconic patterns to showcase feminine strength (Mary Katrantzou, n.d.; TCM Magazine, 2014). As for Adidas by Stella McCartney, through which Stella has collaborated with Adidas for more than a decade, her uncompromising convictions are found throughout the Spring–Summer 2019 collection, which is all about bringing sports, wellness, and—most importantly—sustainability into women's lives (Adidas America Inc, 2019a). The collection has a jacket, which uses both flat-knit construction and super lightweight woven fabric for the sleeves, and other stretchy Primeknit items, which use sustainable materials that partly move with the wearer's body like a second skin (Fig. 2.11) (Stella Mccartney, 2019). These collaborations have helped Adidas to create energy for their brand and caused others to give Adidas another look into their products.

2.4 Fashion brands: activewear collections

The shift toward a more active lifestyle has led to the continuous growth of the global sportswear market. This has attracted fashion labels like Zara and H&M to follow the steps of Nike and Adidas and take part in the ever rapidly increasing fashion-meets-fitness trend (The Business of Fashion, 2014). However, a fashion brand needs to have creditability in establishing and building their label in performance wear. Consumers have to find them credible enough to provide garments with the same level of function as for example, Nike, even if the vast majority do not plan to be wearing these apparel for physical exercise (The Business of Fashion, 2014). Although Zara and H&M do not offer the same technologically advanced materials as the traditional functional brands; they still have a price advantage (Reuters, 2014).

2.4.1 Zara

Zara is a Spanish company founded in 1975 and one of the best-known fashion brands globally (Business Insider, 2019). The company is owned by Inditex, the largest apparel retailer worldwide (The Economist, 2011). Zara is best known for its inexpensive and trendy pieces (South China Morning Post, 2016). They launched their first sportswear collection in March 2016 (Fashion United, 2016). In terms of their brand and trend, consumer research has shown that this line of activewear is destined to benefit modern women who have on the go lifestyles by promoting chic and modern aesthetics (Tara Ellis, 2019). Zara has four categories of wear: lightweight apparel such as leggings, swimwear, monochrome basics such as sweatshirts and finally, loungewear, such as yoga pants (Fashion United, 2016). They also offer mix and match pieces that are fabricated from materials with greater breathability and regulate the body temperature (South China Morning Post, 2016). Flashes of neon color have been applied throughout their activewear collection; see Fig. 2.12 (Fashion United, 2016).

For its Fall/Winter 2016 line, Zara launched a motivational athleisure-inspired gym wear collection, which was all about diversity and versatility and inspired by the Olympics (Fashionisers, 2016). In the collection, there are boxeur jackets and yoga pants, which gently hug the body and make it easier for movement (Fashionisers, 2016). There are also billowy hoodies, boxing gloves, and relaxed shirts (Fashionisers, 2016). Since the collection is more athleisure wear than conventional gym wear, all of the products can be worn outside of the gym and during both day and night. All of the

Figure 2.12 Some of the items in Zara's first sportswear collection: Spring/Summer 2016.
Source: Fashion United (2016).

garments are casually chic, and feature creamy nude and pastel colors except for a yellow hoodie and a metallic jacket, for chic look (Fashionisers, 2016).

Aside from gym wear, the Zara swimwear collections also target those who want to be fashionable but limited to a budget. For instance, their 2019 swimwear/ beachwear collection features eye-popping prints and shimmering colors (Elite Daily, 2019). Apart from swimwear, the line also includes bags, sunglasses, cover-ups, hats, sandals, and even hair scrunchies (Elite Daily, 2019). The collection is well curated with different choices for consumers (e.g., multicolored versus solid, or one-piece versus bikini, etc.) so that there is something for every preference (Elite Daily, 2019).

2.4.2 H&M

One of the longer global retail brands that has been around since 1947 is Swedish based H&M, which offers trendy and stylish fashions for those on a budget (SuccessStory, 2018). To date, H&M is only second to Inditex as the largest clothing retailer in the world. In January 2014, H&M presented its H&M Sport line, which was designed based on feedback from Swedish Olympic athletes and considered to be a real contender to the other sportswear labels (The Business of Fashion, 2014). H&M stated that their line, which targeted a variety of physical activities such as yoga and tennis for both adults and children, has been tested by the Olympic team, and offers shirts that dry quickly and sports bras that are breathable (Reuters, 2014). The limited edition sportswear items were worn by the Swedish Winter Olympic athletes in the Go Gold Collection, campaign which showcased a quick drying jersey, running pants with a snake skin theme, down jackets with duck feathers, and a running vest that can be converted into a dress (Fashion Tape, 2014).

After that, H&M continued to launch different sportswear collections. In 2018, they launched their first activewear collection, which featured sustainable and recycled materials in response to the market demand for environmentally friendly clothing options (Welltodo Global, 2018). This collection, which is called Conscious-Led Activewear Line, includes training, running and yoga tops, sports bras and tights, among others and emphasizes nature in its patterns and colors (Welltodo Global, 2018; PromoStyl, 2018). A new knitting technique is used for the collection to create seamless garments that use less yarn and fabric to avoid waste (Welltodo Global, 2018).

2.5 Lingerie brands: activewear collections

2.5.1 Victoria sport

The largest retailer of women's lingerie in the United States, Victoria's Secret, was founded in 1977 (Katherine, 1986). In 2013, Victoria's Secret offered its first sports bra to the market, which was largely unsuccessful in sales (Mark, 2016). After that, the company started to focus more on performance (Mark, 2016) and launched their current sports and workout apparel line called Victoria Sport in 2016 (Business Insider, 2016). The company has been changing its image from sexy to strong and

Figure 2.13 "Bra-in-a-bra" design of Victoria's Secret.

athletic (Business Insider, 2016). In doing so, they have managed to cater to consumers today who are more geared toward health and wellness (Business Insider, 2016). Their sports bras are available in real sizes and not just general sizes such as small to large. Therefore, they have an international reputation for fit, quality, and style, which has successfully surpassed the marketing pitch of the Nike Expert Session bra fittings (Econsultancy, 2016; News Corp Australia, 2016). The brand also encourages women to maintain fitness regardless of their size by offering its apparel in a large range of sizes (Bon Wells, 2016).

Apart from providing more options in size for its customers (e.g., from 32B to 40DDD for the Max Sport Bra and the Ultra Max Sport Bra series) (Victoria's Secret, 2019), the brand also applies innovative elements to its sportswear line. Take their patented "bra-in-a-bra" technology for example. This technology claims to have unprecedented support (Victoria's Secret, 2019). The bras provide maximum support with underwire bra cups that clip together at the center as well as another layer of fabric that surrounds the inside bra for more compression and control of bounce (Victoria's Secret, 2019a). Wicking fabric is used to keep the wearer cool and dry and thus optimizes her level of comfort (Victoria's Secret, 2019a). The bras also have padded straps for stabilized support and can be easily donned and doffed (Victoria's Secret, 2019a). Fig. 2.13 shows the "bra-in-a-bra" design of Victoria's Secret.

2.5.2 CW-X

Wacoal Holdings Corporation, the largest intimate apparel company in Japan, first offered the CW-X line for the performance sports apparel market in 1991 for Japan and 2002 for the United States (CW-X, 2019a). They market the CW-X line as a pioneering anatomically engineered sports apparel for high performance effects and, to date, has over 50 function and design patents worldwide (CW-X, 2019a). CW-X combines both the scientific human physiology and advanced fabrics and apparel, to produce different performance sports apparel for athletes (CW-X, 2019a). The products are both science-based and anatomically engineered with premium-quality, which

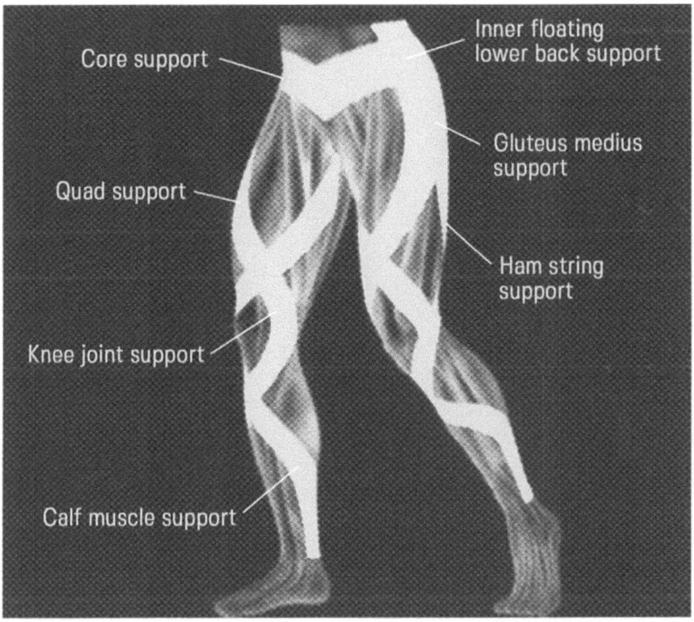

Figure 2.14 Support panel of Endurance Generator Compression Tights with Muscle & Joint EXO-WEB technology.
Source: CW-X (2019a).

help to improve sports or fitness performance (CW-X, 2019a). The patented technologies include, for example, Muscle & Joint EXO-WEB, Joint EXO-WEB, Muscle EXO-WEB, and Sports Bra with SUPPORT-WEB (CW-X, 2019b).

The Muscle & Joint EXO-WEB is applied to the CW-X Endurance Generator Compression Tights, which feature the patented technology to support muscles and joints when the wearer is doing high impact activity (CW-X, 2019c). A support panel is used to stabilize the muscles and joints but does not inhibit movement (CW-X, 2019c), see Fig. 2.14. These tights compress much like kinesiology tape to provide supportive forces onto vulnerable areas such as the hamstring muscles and hip joints (CW-X, 2019c). The company states the benefits of compression from the tights allow for better circulation, which means that the wearer is less prone to muscle fatigue and faces a lower likelihood of damage to the muscles resultant of the support and stabilization of the muscles. Therefore, the wearer experiences enhanced performance and can recover from muscular injuries in a shorter amount of time (CW-X, 2019c).

Another example of the CW-X line is the Stabilyx High Impact Sports Bra, which features the SUPPORT-WEB technology. Mesh webbing, which offers high compression, is constructed as part of the sports bra cup to reduce bounce and increase support of the breasts during high impact workouts (CW-X, 2019d). The bra is constructed of the EXO stretch material that is not only a very breathable material but also wicks moisture so that the wearer feels dry (CW-X, 2019d). Apart from a flat seam construction that prevents chafing and irritation, the bra also has a fully adjustable clasp back

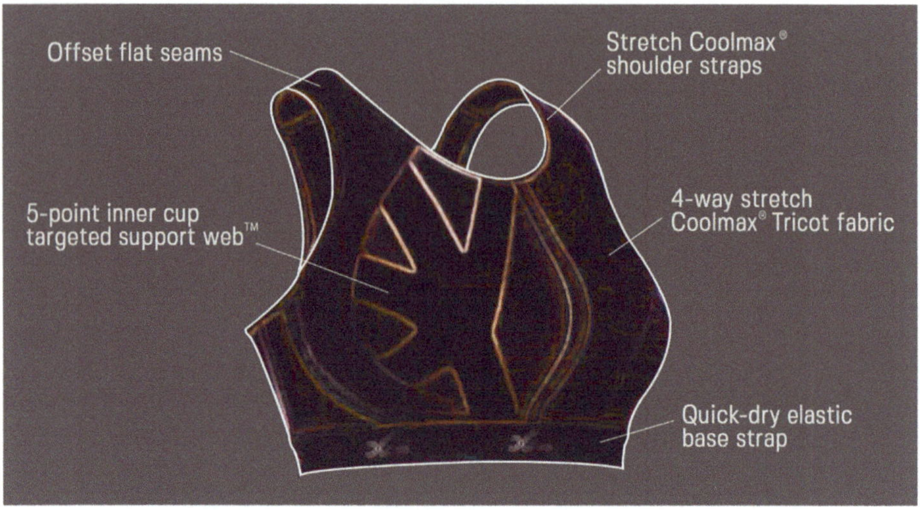

Figure 2.15 Sketch of Stabily x High Impact Sports Bra with SUPPORT-WEBT technology.
Source: CW-X (2019b).

and shoulder straps so that those with larger breasts (e.g., D and DD cups) can also wear the bra; see Fig. 2.15 (CW-X, 2019d).

2.6 Athletic leisurewear trends

Athleisure can be simply defined as a fashion trend, which integrates athletics and leisure so that workout clothing is worn for other purposes such as for school or work other than just physical exercise (Derorah Weinswig, 2012). Athleisure is also known as sportswear chic (South China Morning Post, 2016). Some of the more common athleisure apparel includes yoga pants, sports bras, shorts, and tights.

The rapid growth of athleisure wear is in part attributed to the fashion industry, which can be traced back to the 1970s when fitness emerged as a popular activity (Derorah Weinswig, 2012). As a result, the different labels began to design clothes that appear to be suitable for athletic activities, which were further enhanced by using innovative synthetic materials such as Lycra and spandex (Derorah Weinswig, 2012). In recent years, the increased awareness of health and wellness among the general population and growing trend of sports and fitness involvement have greatly contributed to the demand for activewear, thus resulting in the popularity of athleisure wear (Research & Markets, 2017). Today, consumers seek garments that can take them from one activity to another without having to change. Therefore, athleisure wear is a means of doing so but without sacrificing trendiness or versatility, and keeping the functionality of the apparel at the same time (Dennis Green, 2017).

The unique advantages of activewear have been often touted, including their breathability and ability to dry quickly or resist bacteria and odors, which have accelerated the growth of the global activewear market (Research & Markets, 2017). However, in very competitive markets characterized by the overemphasis on technical aspects, which are often misunderstood, aesthetic considerations may simply push consumers to make their ultimate choice for purchase (Antoine Haincourt, 2019). Therefore, apparel companies take aesthetics into consideration by being creative during the design stages to meet the needs of consumers for not only function but also perceived appeal (Hodges et al., 2018).

On the other hand, product customization and personalization are also becoming more popular, in which the application of new technologies can facilitate their implementation (Hodges et al., 2018). For instance, Vans, an American shoe manufacturer, allows their customers to choose from a wide range of designs to construct their own unique shoe. Nike also gives their customers the opportunity to create a unique shoe by choosing their own color schemes and adding text.

Clothing is now becoming more flexible and personalized, as exemplified by the success of athleisure apparel (Antoine Haincourt, 2019). Athleisure clothing is a mix of different styles that can be practical, aesthetically pleasing, casual, and athletic, and at the same time, is both beautiful to look at and comfortable to wear (Antoine Haincourt, 2019). The silhouettes have also become less formal. The major sports brands, or even ready-to-wear brands, have already expanded their threshold. They are now producing more leisure clothing, while the brands that have conventionally offered only athletic clothing are now incorporating fashion elements into their garments (The Telegraph, 2016). This shows that athleisure is both about aesthetics and about health and well-being.

In fact, athleisure has become so popular that it is now preferred over denim (Derorah Weinswig, 2012). Plus, celebrity endorsements have only added to the athleisure trend (The Telegraph, 2016). These celebrities include singer Beyoncé for Ivy Park, and American rapper Pharrell Williams for Adidas. While some say that the athleisure market is saturated, many feel that there is still room to grow based on industry statistics. Since 2009, the industry has grown to a USD 270 billion market and anticipated by Morgan Stanley to reach USD 350 billion in 2020 (Market Watch, 2018).

2.7 Conclusions

With the growth of athleisure apparel and current trend in spending more time on sporting and physical activities for health and wellness, functional athletic apparel, fashion and lingerie brands are putting more effort into their innovations by using technology and imposing creativity into the garment construction design and process of their sportswear products, in anticipation of boosting their sales in the market. The sportswear products need not only be "functional" or "good looking,' but can be both to meet the needs of consumers. Consumers now have more options when purchasing their workout and exercise gear, whether it is to purely give off a "sporty" vibe or the need for professional sportswear to participate in sporting activities. The products of some brands are selling at a relatively higher price, while some accommodate wearers who are on a budget. The price level difference may be related to the amount of investment on innovation, the type of material used, challenges during production, popularity of the brand, market strategy, etc. Even consumers, who have different needs for sportswear products (i.e., different price levels, functional performances, and aesthetic elements), can find a product with different elements that would allow them to realize their goals and objectives.

References

Adidas (2019). Climachill Tee. Retrieved from: https://www.adidas.com.hk/item/EI6391?locale=en_GB.

Adidas (2019b). Believe This Primeknit LTE Tights. Retrieved from: https://www.adidas.com/us/believe-this-primeknit-lte-tights/DU3852.html.

Adidas America Inc. (2019). Creating Change with Stella Mccartney. Retrieved from: https://www.adidas.com/us/adidas_by_stella_mccartney_story.

Alexandria Sage (2008). Nike Profit Up but Shares Tumble on U.S. Concerns. Retrieved from: https://www.reuters.com/article/us-nike-results/nike-profit-up-but-shares-tumble-on-u-s-concerns-idUSWNAS924120080625.

Alice Morby (2017). Nike unveils Advanced Apparel Exploration clothing designed by data. Retrieved from: https://www.dezeen.com/2017/09/09/nike-unveils-advanced-apparel-exploration-collection-designed-by-data/.

Antoine Haincourt (2019). The Global Sport Economy: Contemporary Issues. Chapter 6 – New Behaviour Trends Among Consumers of Sporting Goods. Routledge.

Bon Wells (2016). Victoria Sport: "Strong is Sexy". Retrieved from: https://medium.com/@Bonnaroo_/victoria-sport-strong-is-sexy-4f148adab9d.

Business Insider (2016). Victoria's Secret just introduced its new Nike killer. Retrieved from: https://www.businessinsider.com/victorias-secret-launches-victoria-sport-2016-7.

Business Insider (2019). Zara is part of the biggest fashion company in the world. Meet the other brands it owns. Retrieved from: https://www.businessinsider.com/zara-inditex-brand-list-2018-10.

Cat Meffan (n.d.). Karlie Kloss Shows Off Her Uber - Toned Physique for NIKE. Retrieved from: https://catmeffan.com/karlie-kloss-shows-off-uber-toned-physique-for-nike-lab-supermodel-fitness/.

Chicago Tribune (2018). The Rise of Lululemon: How America Became A Nation of Yoga Pants. Retrieved from: https://www.chicagotribune.com/business/ct-biz-lululemon-yoga-pants-fashion-20181031-story.html.

CW-X (2019a). The CW-X Story. Retrieved from: https://cw-x.com/about/.

CW-X (2019b). CW-X Patented Support Technology. Retrieved from: https://cw-x.com/technology/.

CW-X (2019c). Endurance Generator Joint & Muscle Support Compression Tight. Retrieved from: https://cw-x.com/product/endurance-generator-tights-women/.

CW-X (2019d). Sports Bra. Retrieved from: https://cw-x.com/product-category/women/sports-bras/.

Dan Howarth (2017). Nike launches lightweight Flyknit sports bra. Retrieved from: https://www.dezeen.com/2017/07/12/nike-fenom-flyknit-lightweight-sports-bra-design/.

Daniel Mitchell (2014). Adidas Launches Climachill Apparel Technology. Retrieved from: https://www.complex.com/sneakers/2014/03/adidas-launches-climachill-apparel-technology.

Dennis Green (2017). Athleisure is Not Just a Trend—It's a Fundamental Shift in How Americans Dress. Retrieved from: http://www.businessinsider.com/athleisure-is-more-than-a-trend-2017-2.

Derorah Weinswig (2012). Athleisure and Sneaker Culture are Changing Mainstream Fashion. Retrieved from: https://www.deborahweinswig.com/fashion/athleisure-and-sneaker-culture-are-changing-mainstream-fashion/.

Econsultancy (2016). Site Comparison: Victoria's Secret vs. Nike Women, the Battle of the Sports Bras. Retrieved from: https://econsultancy.com/site-comparison-victoria-s-secret-vs-nike-women-the-battle-of-the-sports-bras/.

Elite Daily (2019). What's In Zara's 2019 Swimwear Collection? Every Trendy Print You Could Ever Want. Retrieved from: https://www.elitedaily.com/p/whats-in-zaras-2019-swimwear-collection-every-trendy-print-you-could-ever-want-17007799.

END (2019). NikeLab A.A.E. 1.0 Tee. Retrieved from: https://www.endclothing.com/gb/nikelab-aae-1-0-tee-aa0216-215.html.

Fashion Tape (2014). H&M Go Gold Collection. Retrieved from: http://ftape.com/media/hm-go-gold-collection/.

Fashion United (2016). Zara launches debut sportswear collection. Retrieved from: https://fashionunited.uk/news/fashion/zara-launches-debut-sportswear-collection/2016022419535.

Fashionisers (2016). Zara Taps Edie Campbell for Its Athleisure-Inspired Fall 2016 Gymwear Collection. Retrieved from: https://fashionisers.com/2016/08/24/zara-fall-2016-gymwear/.

Fashionista (2014). Why NIKE Is Working with Karlie Kloss. Retrieved from: https://fashionista.com/2014/10/nike-karlie-kloss.

Fast Company (2005). Protect This House. Retrieved from: https://www.fastcompany.com/53752/protect-house.

Female for Life (n.d.). How Activewear Can Help Motivate Your Active Lifestyle. Retrieved from: https://www.femaleforlife.com.au/Exercise-and-Fitness/activewear-motivates-healthy-lifestyle.htm.

Fighter X Fashion (n.d.). Under Armour AllSeasonGear Reversible Shirts. Retrieved from: http://fighterxfashion.com/under-armour-allseasongear-reversible-shirts/.

Forbes (2018). Nike, Already World's Largest Sneaker Maker, Proves It's Still A Growth Company. Retrieved from: https://www.forbes.com/sites/andriacheng/2018/12/20/nike-just-proves-its-still-a-growth-company/#12c161e841f7.

Garage Gym Power (n.d.). Best Workout Clothes for Men & Women: Ultimate Guide with Reviews 2019. Retrieved from: https://garagegympower.com/best-workout-clothes-for-men-and-women/.

Group of BATGMX (2013). Futures for Adidas - A course project for BATGMX 2013. Retrieved from: http://adidasfutures.blogspot.com/2013/04/brand-portfolio.html.

Highsnobiety (2016a). How Nike Used the Olympics to Beat adidas to Market with Flyknit Technology. Retrieved from: https://www.highsnobiety.com/2016/08/04/nike-flyknit-vs-adidas-primeknit/.

Highsnobiety (2016b). Adidas Originals Just Launched a Primeknit Apparel Collection. Retrieved from: https://www.highsnobiety.com/2016/01/09/adidas-originals-primeknit-apparel/.

Hodges, N. J. & Albert, N. (2018). Innovation by design. *Small Business Economics*, 52(3).

House of Coco (2018). From Function To Fashion: The Biggest Sportswear Trends. Retrieved from: https://houseofcoco.net/from-function-to-fashion-the-biggest-sportswear-trends/.

Hyperbeast (2015). Adidas Is Making Primeknit Apparel for the First Time in History. Retrieved from: https://hypebeast.com/2016/1/adidas-designs-primeknit-apparel-for-the-first-time-in-history.

Just-style (2016). Puma Develops Temperature Control Football Jacket. Retrieved from: https://www.just-style.com/news/puma-develops-temperature-control-football-jacket_id127114.aspx.

Katherine, B. (1986). An Elegant Kid's Store Fails. The New York Times Company. Retrieved from: https://www.nytimes.com/1986/12/27/business/an-elegant-kids-store-fails.html.

Knitting Industry (2012). Let the games begin: Nike Flyknit versus adizero Primeknit – in court. Retrieved from: https://www.knittingindustry.com/let-the-games-begin-nike-flyknit-versus-adizero-primeknit-in-court/.

Lululemon Athletica (n.d.). *History*. Retrieved from: https://info.lululemon.com/about/our-story/history.

Lululemon Athletica (n.d.). Fabric + Technology – Luon®. Retrieved from: https://info.lululemon.com/design/fabrics-technology/luon.

Lululemon Athletica (n.d.). Fabric + Technology – Silverescent®. Retrieved from: https://info.lululemon.com/design/fabrics-technology/silverescent.

Lululemon Athletica (n.d.). Fabric + Technology. Retrieved from: https://shop.lululemon.com/designs/_/N-1z13xjmZ1z140xk.

Mark J. Miller (2016). Victoria Sport: Victoria's Secret Looks to Athleisure for Growth. Retrieved from: https://www.brandchannel.com/2016/07/27/victoria-sport-collection-072716/.

Market Watch (2018). Athleisure Wear is Crushing it on the Street. Retrieved from: https://www.marketwatch.com/press-release/athleisure-wear-is-crushing-it-on-the-street-2018-10-10.

Mary Katrantzou (n.d.). Mary Katrantzou X Adidas Originals. Retrieved from: https://www.marykatrantzou.com/project/mary-katrantzou-x-adidas-originals/.

Men's Running (2013). Puma Pure Niightcat Powered Jacket. Retrieved from: https://mensrunninguk.co.uk/gear/puma-pure-nightcat-powered-jacket/.

Michael Angelo Quiroz (2018). A Growth Strategy for Lululemon 2018. Retrieved from: https://medium.com/@michaelangelo_q/a-growth-strategy-for-lululemon-68819680e511.

Mirror (2019). Under Armour launches UA Rush sportswear that RECYCLES energy as you exercise. Retrieved from: https://www.mirror.co.uk/tech/under-armour-launches-ua-rush-15004813.

Nathalie Martin (2015). Fashion & Function: Sportswear as a Statement of Style. Retrieved from: https://medium.com/@NathalieM/fashion-function-sportswear-as-a-statement-of-style-fbc073f1ec04.

News Corp Australia (2016). Victoria's Secret launches Victoria Sport range. Retrieved from: https://www.news.com.au/lifestyle/fashion/designers/victorias-secret-launches-victoria-sport-range/news-story/9f06e2785d517bbdf3e20efbd97438f2.

NIKE, Inc. (2015). NikeLab presents Johanna Schneider Women's Training Collection: Modularity for the Body in Motion. Retrieved from: https://news.nike.com/news/nikelab-presents.

NIKE, Inc. (2017). Annual Report on Form10-K. Retrieved from: https://last10k.com/sec-filings/nke/0000320187-17-000090.htm#s0F28647A0E8708EDE85DB1AD7BD67714.

NIKE, Inc. (2017). A Glimpse into the Future By Way of a T-Shirt. Retrieved from: https://news.nike.com/news/nikelab-advanced-apparel-exploration-aae-1.

NIKE, Inc. (2019). Celebrating the Sports Bra. Retrieved from: https://news.nike.com/news/nike-flyknit-sports-bra.

PromoStyl (2018). Collabs Sport Summer 2018. Retrieved from: https://www.promostyl.com/category/sport-en/sport-brand-collaborations/?lang=en&cn-reloaded=1.

Puma SE (2017). Annual Report 2017 - Forever Faster PUMA. Retrieved from: https://annual-report-2017.puma.com/en/company-overview/product/.

Puma SE (2018a). Culture of Firsts. Retrieved from: https://about.puma.com/en/innovation.

Puma SE (2018b). Puma Innovations. Retrieved from: https://about.puma.com/en/innovation/innovation-landmarks.

Research and Markets (2017). Activewear Market by Product ; and Fashion Brand) and Fabric - Global Opportunity Analysis and Industry Forecast, 2017–2024. Retrieved from: https://www.researchandmarkets.com/reports/4580630/activewear-market-by-product-and-fashion-brand.

Reuters (2014). H&M lines up to compete in booming sportswear market. Retrieved from: https://www.reuters.com/article/olympics-hm/hm-lines-up-to-compete-in-booming-sportswear-market-idUSL6N0JY3CI20140109.

Riley Jones (2013). Know YourTech: Under Armour AllseasonGear. Retrieved from: https://www.complex.com/sneakers/2013/11/know-your-tech-under-armour-allseasongear.

Robert J. Terry (2005). Analysts: Brand crossover risky for Under Armour. *Baltimore Business Journal*. Retrieved from: https://www.bizjournals.com/baltimore/stories/2005/09/26/story6.html.

Robert Lewis (2017). Adidas AG - Germen Company. Retrieved from: https://www.britannica.com/topic/Adidas-AG.

Skemateam7 (2013). Adidas Portfolio Strategy. Retrieved from: https://sportswearbrandstrategy.wordpress.com/2013/11/18/adidas-portfolio-strategy/.

So Sasha (2014). 2014 Nike Women Innovation Summit in NYC. Retrieved from: https://www.sosasha.com/2014/11/2014-nike-women-innovation-summit-in-nyc/.

South China Morning Post (2016). The hottest athleisure wear in Hong Kong, from Alala to Zara Sport. Retrieved from: https://www.scmp.com/lifestyle/fashion-luxury/article/1987099/hottest-athleisure-wear-hong-kong-alala-zara-sport.

Sports Unlimited (n.d.). Under Armour All Season Gear. Retrieved from: https://www.sportsunlimitedinc.com/allseasongear.html.

St. James Press (2004). Under Armour Performance Apparel History. International Directory of Company Histories, Vol.61. Retrieved from: http://www.fundinguniverse.com/company-histories/under-armour-performance-apparel-history/.

Statista (2018). Sportswear/Sporting goods companies ranked by worldwide revenue in 2017 (in billion U.S. dollars)* In Statista. Retrieved from: https://www-statista-com.ezproxy.lb.polyu.edu.hk/statistics/241885/sporting-goods--sportswear-companies-revenue-worldwide/.

Statista (2018). Sales of major apparel manufacturers and retailers worldwide in 2018 (in billion U.S. dollars)* In Statista. Retrieved from: https://www-statista-com.ezproxy.lb.polyu.edu.hk/statistics/242114/sales-of-the-leading-10-apparel-retailers-worldwide/.

Stella Mccartney (2019). Adidas by Stella Mccartney. Retrieved from: https://www.stellamccartney.com/hk/online/adidas-by-stella-mccartney/women/clothing.

SuccessStory (2018). The Premier Leader of Global Fast Fashion and Conscious Style: H&M Story. Retrieved from: https://successstory.com/companies/fabulous-fashion-for-all-and-fashion-with-a-conscience-profile-on-hm.

Tara Ellis (2019). A Conceptual Introduction of Sportswear to Zara Retailers. Retrieved from: http://tara-ellis.com/zara-sport.

TCM Magazine (2014). Mary Katrantzou X Adidas: Sports Luxe Collection. Retrieved from: http://www.theclothesmaiden.com/mary-katrantzou-x-adidas-sports-luxe-collection/.

The Business of Fashion (2014). For the Activewear Market, There's No Way But Up. Retrieved from: https://www.businessoffashion.com/articles/intelligence/activewear-lululemon-nike-hm-sweaty-betty.

The Business of Fashion (2015). Why Are Sportswear Giants Nike and Adidas Embracing Fashion? Retrieved from: https://www.businessoffashion.com/articles/intelligence/nike-adidas-sportswear-fashion.

The Economist (2011). Global Stretch - When will Zara hit its limits? Retrieved from: https://www.economist.com/business/2011/03/10/global-stretch.

The Telegraph (2016). The Rise of Athleisure: How the High Street is Cashing in on Office-to-gym Wear. Retrieved from: https://www.telegraph.co.uk/business/2016/09/17/the-rise-of-athleisure-how-the-high-street-is-cashing-in-on-offi/.

The Unique Group (2016). How The Sportswear Industry Balances Functionality and Style. Retrieved from: http://www.theuniquegroup.com/sportswear-industry-balances-functionality-style/.

Under Armour (2019a). Our Mission Is Clear: Under Armour Makes You Better. Retrieved from: https://www.underarmour.co.uk/en-gb/DNA.html.

Under Armour (2019b). Our Story. Retrieved from: https://about.underarmour.com/brand/our-story.

Under Armour (2019c). AllSeasonGear®. Retrieved from: https://www.underarmour.co.uk/en-gb/technologies/allseasongear/.

Under Armour (2019d). Women's UA RUSH Short Sleeve. Retrieved from: https://www.underarmour.co.uk/en-gb/womens-ua-rush-short-sleeve/1332468.html.

Under Armour (2019e). Rush Collection. Retrieved from: https://www.underarmour.co.uk/en-gb/rush-infrared-technology-athletic-apparel/.

Victoria's Secret (2019). Explore Our Sports Bra Collections. Retrieved from: https://www.victoriassecret.com/vs/sport/shop-by-support.

Victoria's Secret (2019). Victoria Sport - Incredible Knockout Ultra Max by Victoria Sport Front-Close Sport Bra. Retrieved from: https://www.victoriassecret.com/vs/sport/shop-by-support/victoria-sport-knockout-by-victoria-sport-front-close-sport-bra-4606f94a-a907-43d2-b776-d944a0b9575b?stackId=8a24cbea-43e0-470e-9abc-22ba9702913b.

Vogue (2017). Will Nike's New Flyknit Bra Change Your Life — Or at Least Your Workouts? Retrieved from: https://www.vogue.com/article/nike-fenom-flyknit-bra.

Wear Testers (2019). Under Armour Rolls Out New RUSH Performance Apparel. Retrieved from: https://weartesters.com/under-armour-rolls-out-new-rush-performance-apparel/.

Welltodo Global (2018). H&M Launches Its First Conscious-Led Activewear Line. Retrieved from: https://www.welltodoglobal.com/hm-launches-its-first-conscious-led-activewear-line/.

Textile fibers and fabrics

3

Joanne Yip, Wing-Yu Chan
Institute of Textiles and Clothing, The Hong Kong Polytechnic University, Hung Hom, Hong Kong

3.1 Introduction

The definition of activewear has evolved over time. In the past, activewear was simply defined as apparel worn during sports activity and hence, even footwear worn for specific activities such as running and hiking, was categorized as activewear (Zurcher Wray & Nelson Hodges, 2008). With changes in fashion trends, the current definition of activewear also includes leisure and casual daytime wear (Ko et al., 2012; O'Sullivan, Hanlon, Spaaij, & Westerbeek, 2017). In general, the definition of activewear involves functionality, comfort and safety to meet the needs of wearers (Hayes & Venkatraman, 2016). Different types of activewear have different design requirements and end products. For example, cycling apparel needs to be smooth and stretchable since it comes in close contact with the skin and provides compression on the body, but at the same time, wicks body moisture for a cooling feeling. However, outdoor winter sportswear needs to maintain thermal insulation and yet also be wind-proof, rain-proof and snow-proof.

The different properties of activewear, such as wicking moisture and providing thermal insulation, are closely related to the materials used in the products. Traditionally, natural fibers such as cotton and wool have been more popular in sport clothing because of their high moisture regain. After the development of fibers with time, synthetic fibers like polyester, nylon and elastane have replaced natural fibers in activewear. Although they are hydrophobic in nature, their properties can be modified because they are moldable, which means that there are many different types of innovative materials that are fabricated from synthetics. These materials provide different features in activewear for different purposes. Therefore, they play an important role in the design and development of sports and leisure wear. In this chapter, the requirements of high-functional and high-performance materials that meet the different consumer and market demands for activewear will be discussed. In addition, an overview of the most popular and innovative types of materials used to meet the needs of consumers and the market in activewear will also be provided.

3.2 Materials requirements of activewear

In the earlier days, the main purpose of activewear was to protect wearers from environmental elements. With time and changes in needs, the requirements for activewear focus not only on just protection but also comfort and different added-value functions, such as the ability to kill or inhibit bacteria. Table 3.1 shows the general criteria for activewear today.

Latest Material and Technological Developments for Activewear. http://dx.doi.org/10.1016/B978-0-12-819492-8.00003-X

Table 3.1 General criteria for activewear.

Criterion related to comfort	Added-value function
Breathable	Bacterial resistant
Cool-touch feeling	Odor resistant
Light in weight	High durability
Moisture permeable	Rain-proof
Smooth	Snow-proof
Soft	Repels soil
Stretchable	Ultraviolet resistant
Dissipates/insulates heat	Water/Wind proof

There are different types of perceived comfort, including thermal-physiological, sensorial, and psychological comfort (Raccuglia, Sales, Heyde, Havenith, & Hodder, 2018). As for the material requirements, they are more related to thermal-physiological and sensorial comfort. Thermal-physiological comfort is related to the weather conditions and the microclimate between the body and clothing. In hot weather, activewear should be able to transfer body heat and moisture out in order to keep the skin dry and contribute to bodily thermoregulation during exercise. Thermal conductivity, moisture wicking, and breathability are the key properties that regulate heat and moisture (Barker, 2002; Jun, Park, Shim, & Kang, 2009; Matusiak, 2010; Parsons, 2014). These properties can improve the wear comfort, but cannot provide a direct cool feeling for wearers during hot summers or strenuous exercise. Cooling materials are therefore more commonly found in activewear. In contrast, winter sports or activities require activewear to have at least thermal insulation ability. At the same time, the activewear has to allow the evaporation of body moisture because accumulated sweat between the skin and apparel will reduce the thermal insulation ability of the garment and thus, the wearer will feel uncomfortable or even stand the risk of hypothermia in cold conditions (Jenkins, 2018).

Apart from thermal comfort, sensorial comfort is also necessary for activewear. Sensorial comfort is the feeling of the fabric against the skin which results from the interaction between the skin and fabric. This feeling comes from the fabric structure, as well as the mechanical and surface properties of the fabric (Uren & Okur, 2019). Sensorial comfort can be further described as sensory factors that cause tactile, moisture, thermal, and pressure comfort (Classen, 2018). Moisture and thermal comfort are feelings that are related to temperature and dampness, namely being wet, damp, cold, hot, etc. They are also related to thermal-physiological comfort or temperature fluctuations of the body. Tactile comfort is related to the pain receptors on the skin so they involve unpleasant feelings such as feeling itchy or scratchy. Pressure comfort is related to the pressure receptors in the skin and involves feelings such as lightweight and heavy. Therefore, sensorial comfort from activewear requires the fabric to wick away moisture to keep the body dry (moisture comfort), or insulate the body against the cold so that it is warm (thermal comfort), or the fabric surface has to be smooth and soft (tactile comfort), or stretchable and light in weight (pressure comfort) to minimize the pressure on the skin during different activities.

In addition to the wear comfort, the functions of activewear are also important. For example, for an activewear, antibacterial and odor resistant are the most common requirements since the space between fibers provides a warm and humid environment for the growth of microorganisms that cause odor and bacteria during and after sweating. In addition, protection against wind, rain and snow in extreme weather conditions are also demanded of activewear, such as skiing, snowboarding and sledding garments (Kanjana & Nalankilli, 2018). Activewear that is worn for recreational outdoor activities that require long hours in different environments such as hiking, fishing and hunting need to be UV resistant, and soil- and water-repellent to protect wearers from their surroundings. In order for activewear to provide continuous protection, the durability of the fabrication materials is also a basic requirement.

3.2.1 Common types of fibers

Traditionally, cotton and wool fibers are the most commonly used types of fibers in activewear. They are hydrophilic which means that they are able to absorb sweat during exercise. Also, cotton provides a soft hand feel while wool is a good insulator to keep the wearer warm. However, due to their high moisture regain, it is difficult to transfer perspiration away from the body or allow the sweat to evaporate. Therefore, the large amount of moisture that remains on the fabric may cause hot and sticky feelings (Havenith et al., 2013; Wang et al., 2019).

Apart from natural fibers, regenerated fibers, which are produced by dissolving natural cellulosic materials and subsequently regenerating the material through extrusion and precipitation, are also commonly found in the activewear market. Since they are created from cellulosic materials such as wood pulp and cotton waste, most of them are hydrophilic in nature. Viscose, lyocell, and modal are usually used for activewear due to their soft hand feel which is similar to cotton. When compared to natural fibers, they provide a smoother surface, allow better moisture management and are more durable (Harlin, Jussila, & Ilen, 2019).

In contrast to natural and regenerated fibers, most synthetic fibers are hydrophobic which means that they resist moisture. An example is polyester, which is frequently used for activewear due to its excellent resistance to water and stains. Polyester does not absorb moisture so it is mostly used to wick away moisture. In addition, it is a strong fiber with high tenacity and good durability and hence, an ideal material for outdoor activewear that protects wearers from wet and damp environments. Apart from its hydrophobicity, polyester is a thermoplastic. Thermoplastics are polymer materials softened at a certain elevated temperature that harden after cooling. The properties of polyester can change when the shape is changed. For instance, polyester itself is not a good insulator. If it is created as a hollow fiber or in a crimped shape, polyester can trap still air inside and insulate body heat. Other synthetic fibers that are appropriate for activewear are nylon, polypropylene, acrylic, and elastane (Shishoo, 2015). However, they are nonbiodegradable, so their massive usage will severely affect the environment. As a result, natural materials are indispensable and contribute to sustainable development.

Nowadays, the demand for activewear is also about versatility. The most basic means of creating activewear to provide a variety of different functions is to blend different types of fibers together. That has led to the development and launch of many innovative types of fibers in the market to meet the different needs of activewear.

3.2.2 Innovative types of fibers

3.2.2.1 Ultra-high molecular weight polyethylene

Ultra-high molecular weight polyethylene (UHMWPE) is a type of polyolefin that is composed of very long chains of polyethylene with a very high percentage of parallel orientation and high levels of crystallinity (Fibre-line, 2019). UHMWPE fibers are produced by gel spinning or dry-wet spinning, which imparts high strength into the fibers. The fibers are drawn into a spinneret during the spinning process, then cooled in water, and stretched to form long oriented molecular chains. There are now primarily two manufacturers that are producing UHMWPE fibers: Honeywell International Inc. and DSM.

The UHMWPE fiber developed by Honeywell International Inc. is called Spectra fiber, which is light in weight, high in strength, resists abrasion, and durable. This gel-spun UHMWPE is a super strong fiber which is 15 times stronger than steel and 30%–40% stronger than aramid (Honeywell International Inc, 2018). However, it is lighter than most traditional or composite fibers such as nylon, polyester, and glass fibers. Fig. 3.1 is Spectra fibers and their applications. These exceptional properties of Spectra have led to their use for outdoor activewear, such as hiking and biking clothes.

The UHMWPE fiber offered by DSM is called the Dyneema fiber. As with the Spectra fiber, the Dyneema fiber has extremely long molecular chains which can transfer load more effectively but is light in weight at the same time. It is up to 15 times stronger than steel and 40% stronger than aramids. Dyneema can float on water and is resistant to moisture, UV and chemicals (DSM Dyneema, inpress). This fiber has been

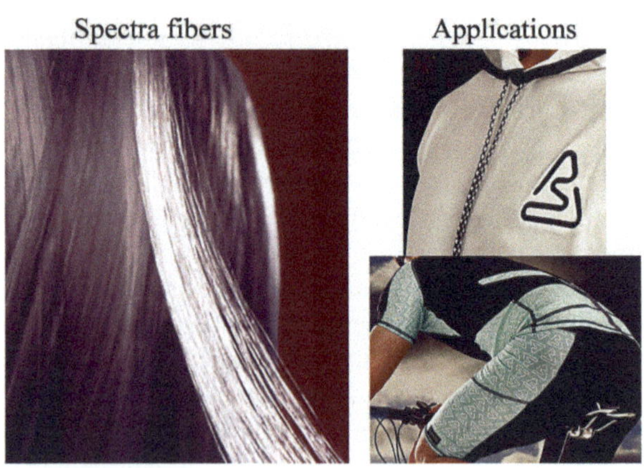

Figure 3.1 Spectra fibers and their applications.
Source: Modified from Honeywell International Inc (2018).

widely used in all-season jackets to withstand all weather conditions and is extremely light in weight. In 2015, Dyneema was also applied to fabricate cycling shorts. The Dyneema reinforced bib-shorts improve the level of protection in the event of more serious crashes by 50% or more without loss of performance or wear comfort. It has been already worn by Team Sunweb-Giant (formerly GIANT-Alpecin team), a German cycling team, and demonstrated its protective value for the team members when they crashed in 2016 (DSM, 2016).

3.2.2.2 Jade fiber

Jade is an ornamental mineral which is well known for its instant cooling touch. Jade also contains trace elements such as zinc, magnesium, copper, selenium, chromium and cobalt. This gemstone is believed to have health benefits as it promotes circulation and increases metabolism (Yu et al., 2015). There are therefore different manufacturing processes that infuse jade particles into textile fibers, yarns, or fabrics. Jade fibers are also normally found in polyester blends. The manufacturing of jade polyester fibers involves the extraction of jade and other mineral materials and grinding them into a nanoscale particle size. They are then mixed with liquid polyester during the melt spinning process. The cooling effect of jade fabric is activated when it comes into contact with body heat, and thus the skin feels cool and dry (Cui, Zhu, Li, Wang, & Liu, 2012).

A numbers of textile companies produce jade fibers, yarns or fabric. One such a company is Qingdao BetterTex Fiber Co., Ltd who developed ice jade which provides both a cooling touch and quickly absorbs and releases perspiration due to the special cross section of the ice jade fibers. The heat transfer rate of ice jade is five times more than that of ordinary polyester fibers (Qingdao Bettertex Fibre, inpress). Neshin Spinning Co., Ltd. developed a yarn called cool jade which is made of amphibole, a kind of jade in Taiwan. Cool jade has very high density and the multi-open-cell structure of the jade particles can contain a higher volume of water and provide an excellent cooling effect (Neshin Spinning, inpress). Performance Sport Limited produced the CoolJade fabric which is infused with recycled jade and provides ultra-wicking that can reduce the skin temperature up to 10°F (Performance Sport Limited, 2019). Chen et al. (2015) also applied and evaluated the performance of polyester jade and Tencel Sun in knitted fabrics. Tencel Sun is a UV protection fiber developed by Lenzing AG, Austria. It is comprised Tencel lyocell fibers with permanently integrated mineral-based pigments. Therefore, polyester jade/Tencel Sun knitted fabrics not only provide a cool feeling, but also has UV protection, antimicrobial, moisture absorption, air permeability, quick drying, and elastic recovery properties. The resultant apparel fabric would meet the demand for comfortable and functional activewear.

3.2.2.3 Synthetic insulation

Natural feathers or down is regarded as a highly efficient insulation material. They are commonly used to fill insulated clothing such as snow and ski wear. However, natural down is less effective in maintaining warmth in damp environments due to their cohesion when wet (Bonser & Dawson, 1999). After repeated washings, the thermal insulation of the down will be gradually reduced. Therefore, synthetic insulation is now more popular in the activewear market. For example, the Climashield brand that is under HarVest Consumer Insulation, LLC offers different types of insulation products such as outerwear, gloves and footwear. Their mission is to provide products that offer thermally-efficient insulation for a long time which they call durable warmth. The technology behind Climashield is to use continuous filaments and hollow fibers to trap more still air for better insulation than the use of traditional staple fibers. The long filaments allow recovery after stretching, compressing or washing the garment so that the filling would not be flattened or split and thus, can provide long-lasting insulation throughout the product life (Harvest Consumer Insulation, 2015a; Harvest Consumer Insulation, 2015b). Fig. 3.2 illustrates the properties of the Climashield continuous filament.

Apart from Climashield, PrimaLoft is another type of innovative synthetic insulation offered by PrimaLoft, Inc. They use recycled microfibers which maintain body temperature and conserve energy. The ultrafine fibers form a number of air pockets that trap air to prevent the dissipation of body heat. Also, PrimaLoft is water resistant, so the air pockets will not be compromised in wet conditions and can retain their insulating properties. A water repellent finishing is applied on the fibers which creates high surface tension that prevents the infiltration of water and helps the fibers stay dry. Furthermore, the small diameter of the fibers facilitates a soft hand feel that mimics the softness and compressibility of natural down and provides air and water vapor permeabilities to enhance wear comfort (Patagonia. PrimaLoft, inpress). PrimaLoft Gold Insulation, Silver Insulation and Black Insulation are the three series of the synthetic downs of PrimaLoft, Inc. They contain different percentages of PrimaLoft fibers to meet the different requirements of customers. Gold Insulation is designed for people

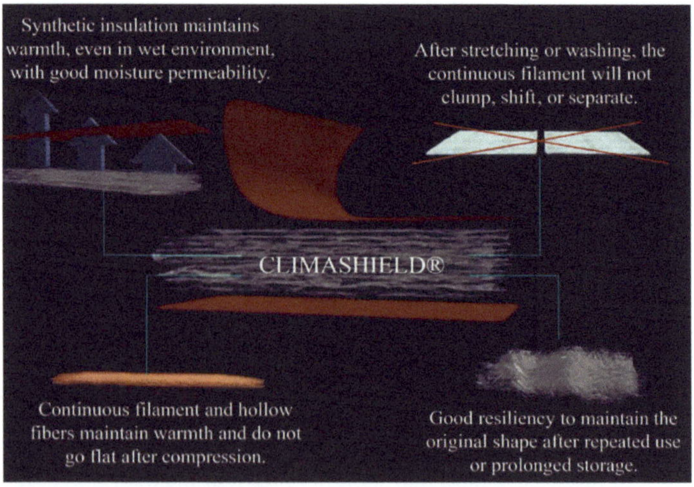

Figure 3.2 Properties of Climashield continuous filament.
Source: Modified from Harvest Consumer Insulation (2015a) and Harvest Consumer Insulation (2015b).

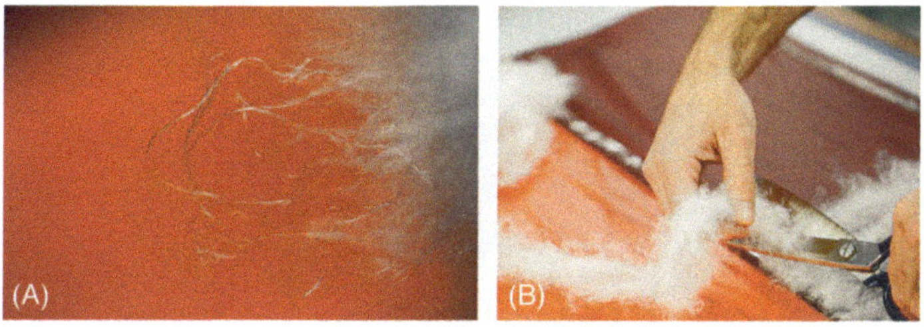

Figure 3.3 (A) Spiral blasted filaments and (B) PlumaFill fluff.
Source: Modified from Patagonia (2018) and Gonzalez (2017).

who lead active lifestyles, even in the most challenging and treacherous weather conditions, Silver Insulation has less goose down and more synthetic down, while Black Insulation is more suitable for everyday wear because it has an equal amount of both types of downs (PrimaLoft, 2018).

PlumaFill is another new synthetic insulation developed by Patagonia. Their inspiration came from the problems of natural and synthetic downs. Natural down loses its insulating properties while synthetic down cannot provide the same amount of warmth, lightness or compressibility as natural down (Gonzalez, 2017). PlumaFill is made of microfilaments that spiral a central fiber and provides a similar look, hand feel and properties as natural down. It is highly compressible and 40% warmer than other types of insulations. Also, after repeated washings, PlumaFill hardly loses its loft (Patagonia, 2018). Therefore, PlumaFill is used to produce the lightweight, warm, and durable Micro Puff jacket for winter or outdoor activities. Fig. 3.3 shows the spiral blasted filaments and the PlumaFill fluffy offshoots.

Nau International Inc. also fabricated the 37.5 Technology (formerly Cocona) insulation which is an alternative to goose down (Nau International, 2016). This technology uses activated carbon derived from coconut husks blended with recycled polyester. The carbon is dark in color which can absorb more heat while the blended polyester fibers increase the surface area to allow quick drying and odor resistance. Therefore, the synthetic insulation developed using 37.5 Technology is able to retain warmth, dries fasterm and resists odors so that it is ideal for cold-weather jackets.

3.2.2.4 Hygroscopic synthetic fibers

Synthetic fibers are mostly hydrophobic which means that they resist moisture and therefore cause increase bodily humidity when sweating due to the low rate of moisture absorption. In order to address this problem, Coolmax, a brand name for a series of moisture-wicking fabrics was developed by Invista (formerly Dupont Textiles and Interiors) 30 years ago. They use polyester fibers with a special cross-section so that the grooves run lengthwise along the fibers. They are manufactured in either a tetra channel or hex channel style. The spaced channels create capillary action that wicks moisture through the core and out to the environment for better evaporation of sweat (Harlin et al., 2019). Fig. 3.4 shows a fiber cross-section of Coolmax fabric.

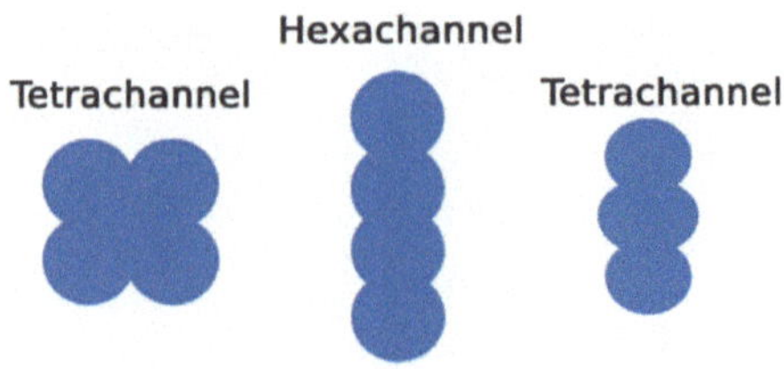

Figure 3.4 Fiber cross-section of Coolmax fabric.
Source: Modified from The LYCRA Company (inpress).

In 2018, Bonaldi et al. (2018) patented a polyamide fiber with improved comfort management. They stated that highly hygroscopic fibers increase the drying time of a garment, which leads the wearer to feel cold after perspiration, so the garment that is soaked with sweat loses its thermal insulation. Therefore, hygroscopic fibers should have reasonable hydrophilicity, and high wicking and quick drying abilities to maintain wear comfort which is achieved with their patented polyamide fiber. Other similar patents have been filed, for example, the nylon fiber patented by Bonaldi et al. (2018) has at least two coalescent centers and five equally dimensioned oblong lobes. Each coalescent center connects with at least three equally dimensioned oblong lobes. The specific multilobal cross-sections of the nylon fiber consists of special micro and continuous channels lengthwise of the fibers which increase the capillarity and surface area. They can increase water absorption, spreading and drying rates which would improve moisture management and wear comfort. Table 3.2 are examples of multilobal cross-section nylon fibers.

3.2.2.5 Cotton-like polyester yarns

Cotton fibers have a nonuniform structure which is flat and twisted like a ribbon. These fibers also have a hollow and kidney-shaped cross-section, which provides more open spaces between the fibers and is the reason for their soft, breathable, and water absorbent properties. In considering the high rates of consumption of cotton fibers and therefore a shortage problem of cotton, there are cotton-like polyester yarns that can be used to simulate the properties of natural cotton fibers.

Murata Machinery, Ltd. developed VORTEX yarn which is created through their unique VORTEX spinning method (Murata Machinery, inpress) so that the center of the yarn is not twisted but one end of each fiber is fastened in it while the other end of the fiber is twisted to form the outer layer of yarn. This is a new concept to spin staple yarn with compressed air. Employing this technology, each fiber is spun and secured in the internal layer of yarn and hence, the sliding of the fiber can be prevented and produce a stable yarn with low hairiness. This structure can be applied in all types of materials. The yarn construction principle is the same for any type of fiber used. However, it is especially suitable for 100% polyester fibers since it mitigates the pilling of 100% polyester spun yarns as it produces yarn with little hairiness.

Table 3.2 Examples of multilobal cross-section nylon fiber.

Sample		Coalescent center	Lobe number	Angle	Illustration of fiber shape
A		2	5	120°	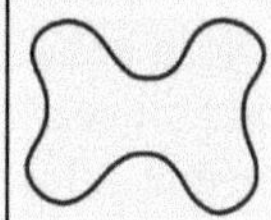
B	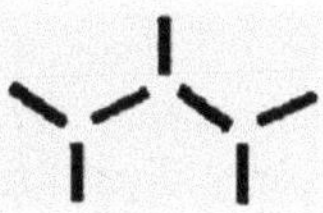	3	7	120°	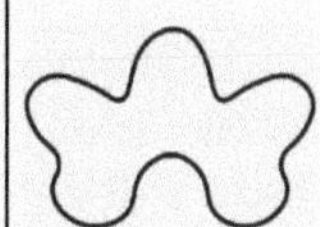
C	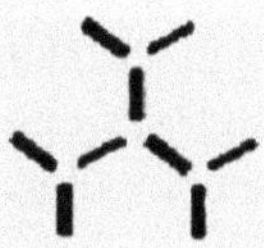	4	9	120°	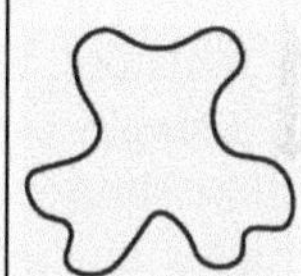
D		4	9	120°	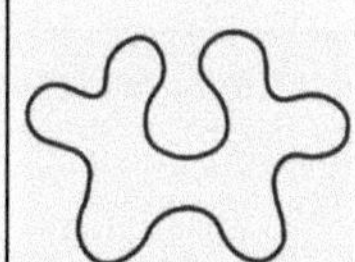
E	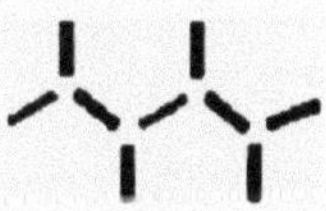	4	9	120°	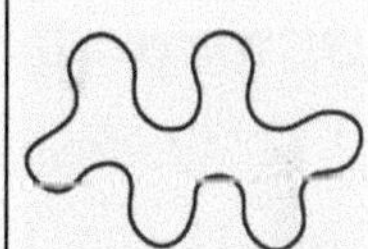
F	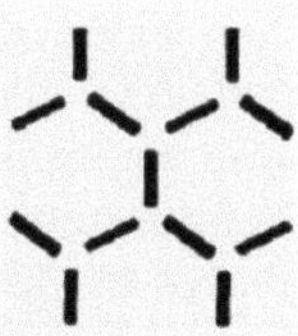	6	13	120°	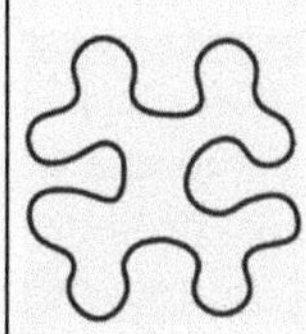
G	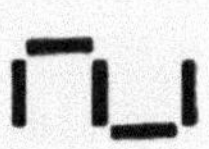	4	5	90°/270°	
H		6	7	90°/270°	
I	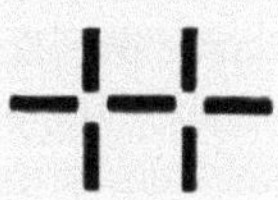	2	7	90°	
J	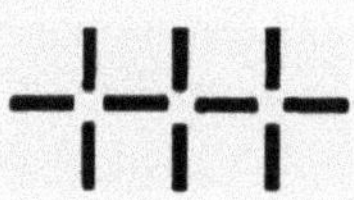	3	10	90°	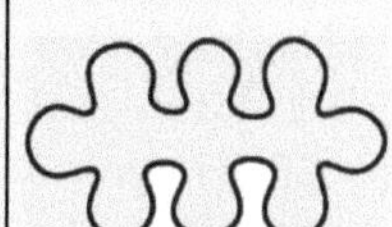

Source: Modified from Bonaldi et al. (2018).

The VORTEX spinning method can also provide a cotton-like texture that is appropriate for sportswear. In comparison to RING spun yarns, also produced by Murata Machinery, Ltd., there are fewer splice points and better uniformity thus resulting in ease of process and better fabric appearance (Murata Machinery, inpress). VORTEX polyester yarn has already been applied in soccer wear which was used by the Gamba Osaka football team due to the cotton-like comfortable touch and water absorbent properties that allow the professional footballers to quickly feel comfortable again after they perspire during training (Murata Machinery, 2014).

Cotna is another type of cotton-like polyester yarn offered by Hyosung TNC (HYOSUNG TNC, 2018). Cotna has a natural hand feel and is an easy care yarn. Cotna has a soft and comfortable touch and absorbs moisture like a natural fiber. At the same time, cotna improves the dimension stability of the fabric, resists piling and manages moisture. Cotna yarn is created by combining low and high shrinkage segments in a single yarn. After the yarn shrinks, the length of the high shrinkage segment will be shortened at a higher ratio and the low shrinkage segment will be bound together with the high shrinkage segment. In this way, the low shrinkage segment will form some loops which stand on the yarn surface. Those loops create a soft touch that mimics the hand feel of natural fibers and provide higher water absorption and a faster drying rate.

3.2.3 Innovative types of fabrics

3.2.3.1 Neoprene

Neoprene or polychloroprene is a synthetic foam-like fabric developed by Dupont (now Invista) in the 1930s (Ünal & Eren, 2018; Grossman, 2017). Neoprene is an organic compound that is composed of carbon and hydrogen atoms. As early as the 1950s, Neoprene was used for surfing wear since it is able to trap air inside itself and hence, provides an insulating barrier between two different temperatures which reduces the rate of thermal conductivity. It is important to ensure that the wearer is not experiencing hypothermia while in extremely cold water. Later, it was found that Neoprene can resist water, salt water, chemicals, and light. Also, it is a synthetic elastomer which is flexible and stretchable. Therefore, more different types of aquatic sportswear such as swim suits, sports bras and diving suits have been subsequently fabricated with Neoprene. Furthermore, Neoprene can be glued or stitched so Neoprene sheets are normally sandwiched between two pieces of outer fabrics, usually polyester or nylon. The combination of Neoprene and elastic fabrics can enhance the flexibility of Neoprene for a better fit and easily donned as a garment. Nowadays, the applications of Neoprene are increasingly wider in the activewear market due to its form-fitting properties. Therefore, Neoprene is not only limited to use in activewear but also includes daily wear like Neoprene skirts, sweatshirts and jackets (Grossman, 2017). Fig. 3.5 shows examples of Neoprene activewear. Recently, Ünal and Eren (Ünal & Eren, 2018) evaluated the comfort properties of Neoprene fabrics used in children tracksuits. They found that there are different advantages when Neoprene is combined with different types of yarns. For example, Neoprene combined with 10% elastane

Figure 3.5 Examples of Neoprene activewear.

knitted fabric provides good stretchability and recovery. It can be applied in the knee or elbow areas of the tracksuit to allow body movement but also protect children when they fall. The Neoprene/cotton fabric has a higher moisture absorption rate and provides a soft hand feel. Therefore, Neoprene is a suitable choice for manufacturing activewear to meet the different needs of customers.

3.2.3.2 Flyknit

Flyknit is an innovative fabric construction technology which is patented by Nike and first introduced in 2012. At the beginning, this technology was applied to sneakers. Nike used a computer-controlled knitting method to shape the entire upper part of the shoe (Daniels, 2018; Jones, 2013; Nike Inc, 2013; NIKE Inc, 2019a). A Stoll flat knitting machine was used to produce Flyknit fabrics. This single piece of seamless knitted fabric can perfectly fit the foot thus resulting in a cosy and almost imperceptible sock-like footwear design (Jones, 2013). Flyknit is made of polyester yarns which are more breathable and flexible than traditional shoes. In addition, the yarns are strategically placed to offer more support in desired areas. For example, nonelastic yarns interlaced on the heel or toe areas provide extra support with lightweight properties (Daniels, 2018). In this way, Flyknit provides the shoe upper with a second-skin feeling, great support and durability. It also eliminates the cutting of various materials of traditional footwear which reduces large amounts of wasted materials. In comparison to the traditional Nike running shoes, Flyknit shoes reduce material waste by 80% (Nike Inc, 2013).

Today, Nike has applied Flyknit to sports bras (NIKE Inc, 2019a). The bras are knitted with the use of ultra-soft nylon/spandex yarn to create different levels of support across a single piece of fabric without the use of wires, pads, stabilizers or elastics.

The band of Flyknit bras is produced with enough thickness to support the breasts. The knit tucks and darts on the bra cups provide conformity and compression to hold the breasts close to the body. The higher neck can avoid undesired jiggles while the straps give support and stretchability. The back area allows the highest flexibility and the front center of the bra is crocheted to allow evaporation of perspiration. Therefore, the multifunctional Flyknit sports bras are fabricated by using different knitted structures of one type of fabric. Also, traditional high-support Nike bras contain up to 41 pieces of materials and 22 seams. Flyknit bras reduce the pieces into 2 panels and a binding that is 30% lighter than other Nike bras. As the result, Flyknit bras can offer maximum wear comfort and support.

3.2.3.3 Flyprint

Apart from Flyknit technology, Nike also developed Flyprint, 3D-printed performance textile fabric applied in footwear. They first applied this technology to produce football cleats (NIKE Inc., 2013). Selective laser sintering (SLS), which is an additive manufacturing method that fuses small particles of materials into a 3D shape by using high power lasers, was used to craft the plate of the cleat. This technology allows more revolutionary cleat patterns and stud placements without the constraints of traditional manufacturing processes. For instance, Nike used an amplified, tri-star stud shape to create optimal traction which is necessary to decelerate and change direction quickly, thus maximizing speed on the turf (Nike Inc, 2014). Now, Nike Flyprint is not only used for the cleat plate but also the shoe upper (Nike Inc, 2018b). Solid deposit modeling (SDM) is used to produce the Flyprint upper, in which a thermoplastic polyurethane (TPU) filament is "unwound from a coil, melted and laid down in layers" (Nike Inc, 2018b). This technology helps to translate the athlete's personal data into textile geometries so that the products are entirely unique and reduces the manufacturing time. Also, the friction resistance between the interlacing of the warp and weft yarns in traditional 2D fabrics poses constraints for the fabric design. Three-dimensional printed fabrics have greater potential for precision-tuned containment due to its fused intersections. The fusing of the materials allows more fabric design possibilities and the printed fabrics are lighter and more breathable than traditional textiles. Flyknit yarns can also thermally bond with Flyprint materials to create an optimal balance of fit and structure and eliminate the glue and stitching process.

3.2.3.4 Biomimetic fibrous membranes

Wang et al. (2018) developed biomimetic fibrous membranes that allow water to be quickly transported and evaporated as smart moisture-wicking fabrics in 2018. They stated that the capillary action contributes to the moisture wicking property of the fabrics that guide perspiration to the outer surface through tiny capillary pores of the fabric microstructure. Coolmax has a large market since it is hydrophobic, and a larger surface area of Coolmax can be quickly dried. However, its application is limited in high humidity environments due to its bidirectional wicking. In order to overcome this limitation, porous membranes that allow directional water transport have been developed to drive water in a single direction by the surface energy gradient or other hierarchical structural features. In their study, an ultra-fast water transport membrane was designed to mimic the networks of vascular plants per Murray's law which determines the thickness of transport network branches. They created a biomimetic micro- and nanofibrous membrane by single-side electrospraying self-synthesized polymer with low surface energy. This is a three-layer fibrous membrane which contains a cellulose acetate (CA-TF) layer covered by a leaf-vein-like TF-629C dip-coated microfibrillated cellulose (MFC-TF) nanofibrous membrane that creates bottom-up hierarchical multibranching networks. Then, a hydrophobic coating on the polylactic acid (PLA-TF) layer is deposited through electrospraying to construct a hydrophobic-to-hydrophilic gradient in the thickness direction. A wetting profile with an asymmetric cross-section is finally developed in the biomimetic porous membranes based on Murray's law. By combining a multibranching porous structure and surface energy gradient, these porous membranes with antigravity directional water transport and quick-dry performance can be applied to activewear, especially those that are meant for heavy exercise or hot and humid weather. Their study demonstrated that the rate of water evaporation of the biomimetic porous membrane is 5.8 and 2.1 times higher than that of cotton and Coolmax fabrics, respectively. Fig. 3.6 illustrates the biomimetic porous membrane in Wang et al. (2018).

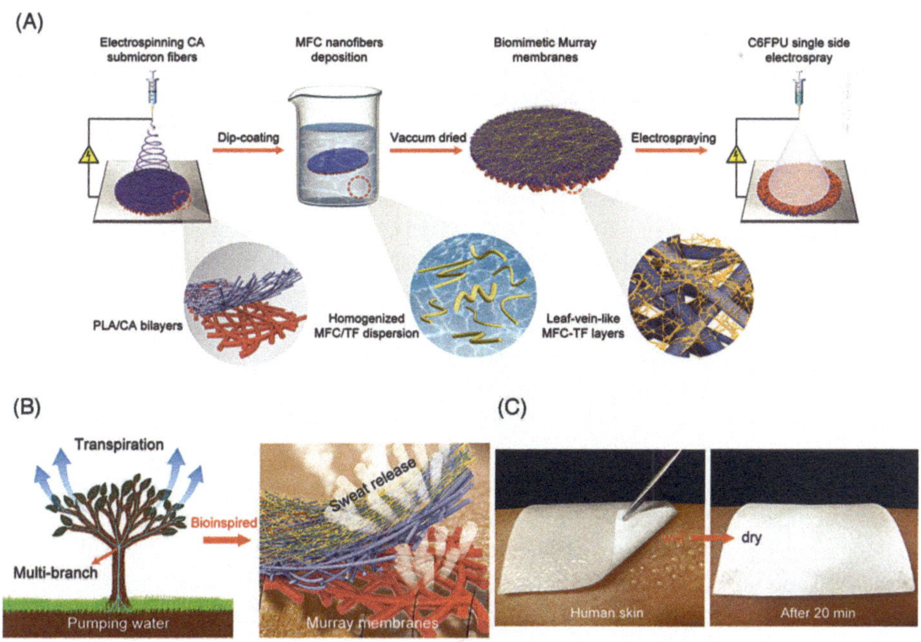

Figure 3.6 Biomimetic porous membrane based on Murray's law. (A) Fabrication processes of the biomimetic Murray membrane, (B) the idea behind the membrane and sweat transport process, and (C) quick dry performance of membrane.
Source: Modified from Wang et al. (2018).

3.2.3.5 TransDRY Technology

TransDRY Technology (Cotton Incorporated, 2019) is developed by Cotton Incorporated in recent years. This technology applies a special water-repellent process on the cotton yarns. TransDRY fabrics are constructed with the treated cotton yarns and the right amount of untreated water absorbent cotton yarns. It allows the quick absorption of sweats and greater movement of moisture from inside to the outside of garment through both horizontal and vertical channels created by the two types of yarns. At the same time, the moisture vapor can be released from the body surface to the atmosphere. With this unique construction of treated and untreated cotton yarns, the overall absorbent capacity of fabric can be reduced so that mimics the synthetic fabrics such as polyester and nylon which would not be over-saturated quickly during exercise. Also, the moisture can be spread over a lager surface area when compared with the traditional cotton fabric which allows quicker evaporation of sweats. Many water-repellent treatments would affect the fabric breathability. TransDRY fabrics combined the absorbent and repellent cotton yarns retain the natural breathability of cotton while it can keep the wearer dry at the same time.

3.2.4 Fabric finishing

3.2.4.1 BioLogic

The Department of Chemical Engineering at the Massachusetts Institute of Technology (MIT) in the USA, Royal College of Art in London, England and New Balance collaborated to invent a new technology called bioLogic (Yao et al., 2015). They used a digital fabrication platform to create hygromorphic biohybrid films on a thin flat substrate which is the deposition of solution-based natural stimuli-responsive material. This is done in a printing process with the use of *Bacillus subtilis* bacterial spores, which is a type of bacteria found in soil (Yao et al., 2015). This technology applies the hygroscopic and biofluorescent behaviors of living cells to the design of biohybrid wearables, which provide multifunctional responses to perspiration. A heterogeneous multilayered structure can be constructed by depositing genetically tractable microbes on a humidity-inert material, thus resulting in biohybrid films that can reversibly change shape and biofluorescence intensity that responds to environmental humidity gradients within a few seconds (Wang et al., 2017). In other words, biohybrid films can react to body heat and moisture and change their shape and form flaps that open to let in air. Wang et al. (2018) also demonstrated the application of bioLogic in the design of a running suit and a fluorescent shoe prototype with bio-flaps. There are air vents that automatically open on the garment which increases air ventilation for cooling down the body and allowing sweat to evaporate. This is an innovative and effective way to produce a mesh structure on activewear. Fig. 3.7 shows the principle of the biohybrid film and its applications.

3.2.4.2 Waterproof breathable fabric

G-Fun Industrial Corporation in Taiwan patented three fabric finishing methods, treatment roller, knife coating or spray head method, that can produce rain-resistant, highly

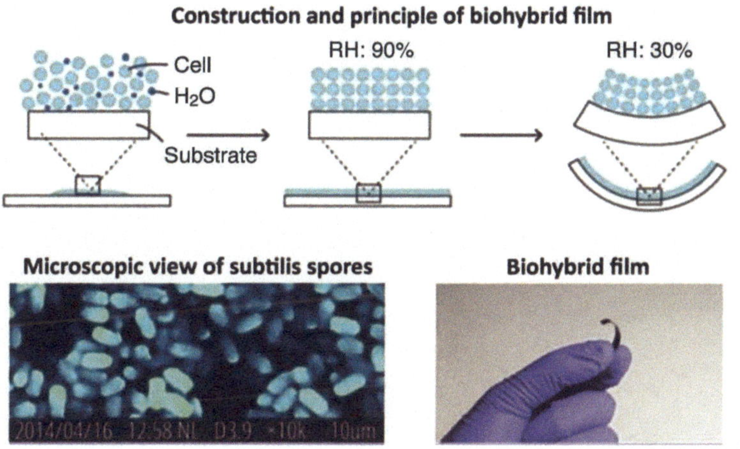

Figure 3.7 Biohybrid film and its applications.
Source: Modified from Yao et al. (2015).

vapor permeable, water-repellent and water-resistant fabrics in 2018 (Cai et al., 2018). A water repellent agent treated on the fabric using their coating technologies can form a thin film on the fabric surface. With the treatment roller method, two types of concave rollers with different sizes are used; one is a back treatment roller and the other is a pressure roller. A knife coater is also placed on the back-treatment-roller to roll back the residual water repellent agent. Then, a pressure roller is used to transfer the water repellent agent and finally a very thin film is formed on the fabric surface by using heat treatment. As for the knife coating and spraying methods, two rollers are used with either a knife coater or a spray head to pass the fabric through the rollers and then the water-repellent agent is coated or sprayed onto the fabric surface. These methods aim to coat a very thin film of strongly water-repellent agent to provide good water-resistance and water-repellence without loss of vapor permeability and wear comfort, which is suitable for manufacturing waterproof outdoor activewear. The traditional water-repellent manufacturing process, which coats synthetic resin by pad-dry-cure method onto the fabric surface, commonly suffers from the problems of insufficient water-resistance and uneven distribution of the water repellent agent on the fabric. Therefore, G-Fun invented this method to improve the water-repellent finishing of fabrics. Fig. 3.8 illustrates the manufacturing process and the result of the waterproof breathable fabric.

3.2.4.3 Herbal extract treatment

The number of research work on herbal encapsulation treatment of textiles has been increasing. Different types of herbs are encapsulated and coated onto the fabric surface in order to provide various functions. For example, Glycyrrhiza glabra (liquorice) can be used to provide an antimicrobial function (Krishnaveni, 2017). Glycyrrhizin, a substance found in the roots of Glycyrrhiza glabra, has an anti-inflammatory property. Therefore, it could be used to heal wounds. It is a recent

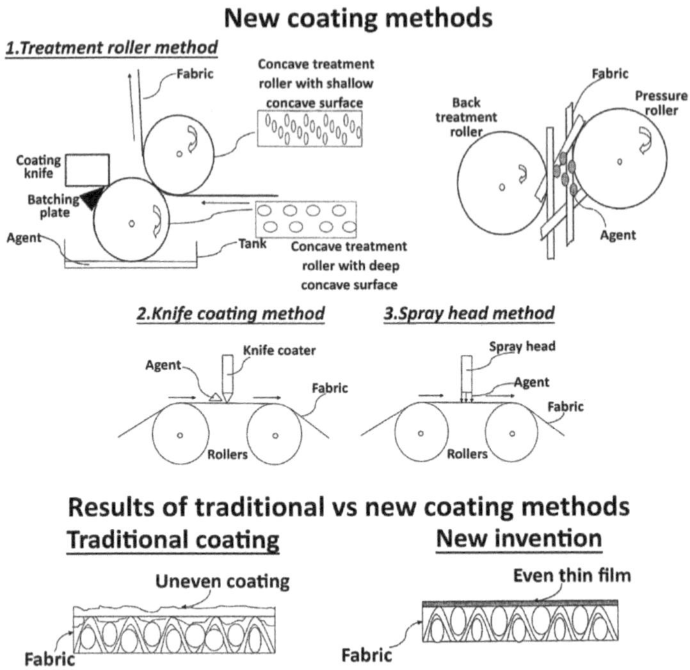

Figure 3.8 Manufacturing process and result of waterproof breathable fabric.
Source: Modified from Cai et al. (2018).

innovation in finishings applied to health-care products. Therefore, textile substrates coated with Glycyrrhiza glabra powder can control the growth of micro-organisms and protect the wearer from environmental hazards. Another example is wild turmeric and holy basil. They are able to facilitate one-way moisture transport capability so as to increase the overall moisture management capability of fabric (Chandrasekaran & Senthilkumar, 2018). Apart from moisture management properties, wild turmeric and holy basil are also known to be good antibacterial herbs which are used in health-care textiles (Mondal & Saha, 2019). In addition, aloe vera extract combined with chitosan has excellent antimicrobial, UV-protection and thermal comfort properties. Aloe vera gel contains approximately 99% water and 1% solid material. Among the solid material, acemannan is the main functional component that has immuno-modulation, antibacterial, antifungal, and antitumor-formation properties. Chitosan is a natural polysaccharide obtained by the alkaline deacetylation of chitin derived from shrimp shells. Those natural agents are inexpensive, biodegradable, nontoxic and safe so that they are suitable for clothing in every way. The direction extraction method is commonly used for extracting herbs. The herbs are first dried and then grinded into a powder, and then the powder is soaked in a solvent such as methanol. After the filtration and evaporation of the solvent, a natural agent can be obtained which is applied onto the fabric by using the pad-dry-cure method. Fig. 3.9 shows the microscopic images of the fabric treated with natural agents.

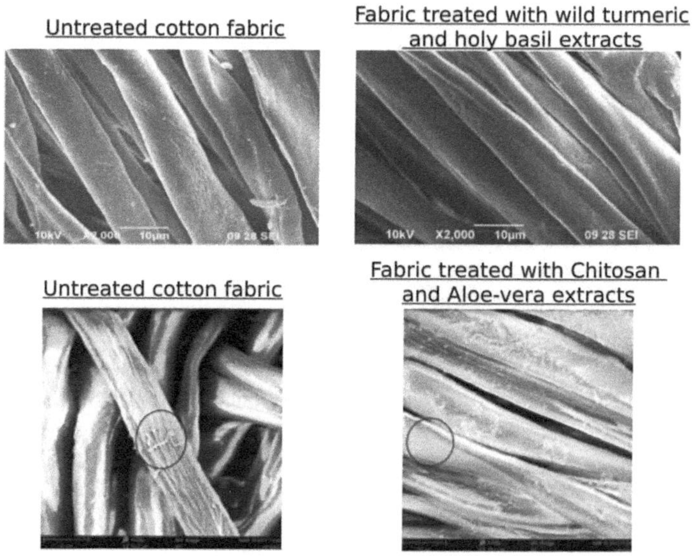

Figure 3.9 Microscopic images of fabric treated with natural agents.
Source: Modified from Chandrasekaran and Senthilkumar (2018) and Mondal and Saha (2019).

3.2.4.4 Finishes for cotton fabric enhancement

There is no doubt that cotton is one of the important fiber types in activewear industry because of its natural hand feel and biodegradability. The only limitations of cotton are the quick moisture saturated nature and poor moisture wicking property which causes the wet feeling during exercise. In order to address these problems and enhance the power of cotton fiber, Cotton Incorporated has developed a variety of categories of cotton materials in their FABRICAST Collection in recent years. For example, their STORM COTTON technology (Cotton Incorporated, 2019) which is a durable water repellent finish for any cotton fabrics or garments which are in knit or woven forms. It helps to repel liquids that protect the wearers from rain and snow, but it would not affect the natural hand feel and breathability of cotton fabrics. It can keep the wearers comfortable, cool and dry anywhere.

WICKING WINDOWS (Cotton Incorporated, 2019) is another technology developed by Cotton Incorporated. The print application of chemistry is applied on the skin-facing side of fabric so as to create the repellent areas and produce "window". Those small absorbent "windows" allows the sweat to be transferred from the skin to the outside of garment. Therefore, it helps to minimize the area of wet fabric that contacts with the skin and keep the skin dry and comfortable during exercise. The printed can be colorless with a simple coverage effect which will only be seen when it becomes wet or with stylized design. There are three methods to apply WICKING WINDOWS technology for different purposes. Single print process of WICKING WINDOWS on the inside of the fabric offers moisture wicking property of garment.

WICKING WINDOWS technology can provide sweat-hiding properties coupled with moisture management by using a double print process. If a phase change material (PCM) incorporate into the WICKING WINDOWS™ print paste, it can also provide temperature regulation and additional cooling properties. Therefore, this technology is versatile that enhances the benefits of cotton fabrics for activewear applications.

3.3 Innovations for sustainability

Sustainability refers to the long-term strategies used to achieve sustainable development. It is the interrelationships among people, resources, the environment, and development which involve the protection and enhancement of the environment (Imperatives, 1987). To achieve the goal of sustainable development, textile industries are making great efforts to find innovative ways of manufacturing that can recycle disposed waste, replace unsustainable resources with green materials and reduce exposure to waste and environment hazards in order to maintain a balance between doing business and protecting the environment.

3.3.1 Recycling

Recycled fibers are becoming more common in the activewear market as many companies are producing fibers by using recycled materials. Unifi is a well-known company that produces recycled fibers. Repreve Bottle Processing Center was established in 2016 and responsible for the Repreve product line of nylon, polyester and staple fibers. To manufacture Repreve, they first sourced recycled plastic bottles and post-industrial waste from industries worldwide. Then, the waste materials were sorted, washed and hacked into flakes. These Repreve flakes were melted and reformulated as Repreve chips. After that, the chips were melted into liquid polymer and extruded through a spinneret to create continuous filaments that are the Repreve fibers. Repreve recycled fibers can be further processed into yarns by spinning. Fig. 3.10 shows the manufacturing process of Repreve yarns. Products such as tops, bottoms, outerwear and socks can all be made from Repreve fabrics (Unifi Inc, 2019a; Unifi Inc, 2019b). This technology creates green products with exceptional performance. Value-added processes like air-jet texturing, yarn covering, solution dyeing and their Tru-technologies can be additionally carried out thus using fewer natural resources to obtain special functions such as moisture management, odor control, thermal comfort, stretchability and water resistance, and reduce the use of water and energy at the same time (Unifi Inc, 2019a).

Apart from the Repreve recycled fibers, there is recycled down offered by Nau International Inc. (Nau International, 2016). Their recycled down factory is in California, USA. To produce the recycled down, they first collect reclaimed duvets and pillows and then remove the down and feathers. After that, the collected down and feathers are sterilized and sorted according to their quality to ensure that the performance of the selected recycled down has the same specifications as virgin down. Fill power, a measurement of down density, is the standard to select recycled down. This means that the number represents how many cubic inches that a single ounce of

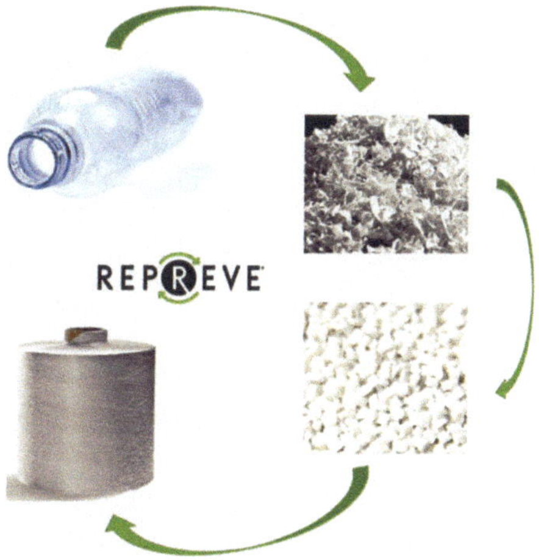

Figure 3.10 Manufacturing process of Repreve yarns.
Source: Modified from Unifi Inc (2019a) and Unifi Inc (2019b).

down will fill. Normally, 650-fill and 700-fill recycled down will be chosen since they have an extremely high warmth-to-weight ratio which is used in high-quality insulated jackets. The recycled down is processed by using the same process as virgin down so it provides the same performance as virgin down and the same care is required. Otherwise, its loft will decrease. Nau also applies water-repellent finishes like Nikwax to the shell fabric to further protect the recycled down from moisture.

3.3.2 Replacing unsustainable materials

Replacing unsustainable materials with green substances is a strategy for sustainability. Activewear industries are committed to offering alternatives to unsustainable materials. One example that has been mentioned here is the production of recycled synthetic insulation which replaces natural down. Goose down is collected by plucking the bird and pulling out the down directly from the goose skin in order to ensure the down quality, which causes wounds on the birds and some may even die as a result. However, the use of synthetic materials is environmentally detrimental since they are nonbiodegradable. To protect both geese and the environment, several manufacturers have produced recycled synthetic insulations like Primaloft and 37.5 Technology insulation which are alternatives to goose down and environmentally friendly.

In addition to recycled synthetic down, Clarino, a division of Kuraray Co., Ltd., is a global leader in the manufacture of nonwoven microfiber material and continues to be the leading innovator in the production of synthetic leather (Formula4 Media, 2016). TIRRENINA is one of the product lines of Clarino that is manufactured with 99% less organic solvent, purely water-based and contains only a minimum amount of polyurethane. TIRRENINA™ is produced without the use of chlorofluorocarbons and free from environmentally harmful additives. Its manufacturing process reduces the usage of water by 70% and CO_2 by 35%. However, it has a similar hand feel and

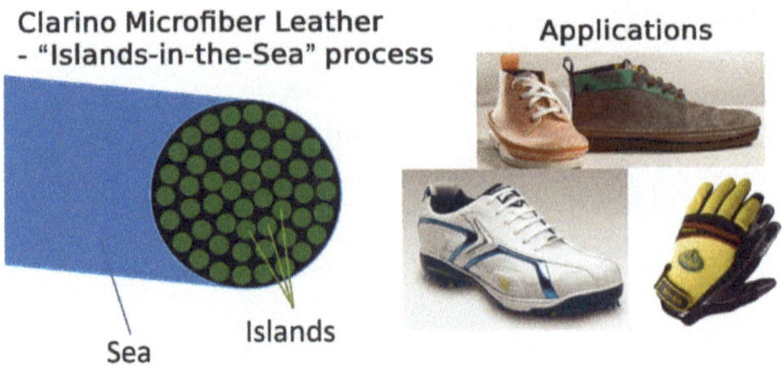

Figure 3.11 Cross-section of Clarino microfiber leather and its applications.
Source: Modified from Formula4 Media (2016) and Kuraray Co (inpress).

cell-like structure as those of animal leather, with ultimate durability, stretchability and recovery. It is three times stronger than genuine leather but only a third of the weight of real leather to achieve equal tear strength. TIRRENINA allows for heat to be dissipated in warm weather and retained in cold weather due to its breathability. Therefore, this material is an environmentally friendly option to replace natural leather. Its characteristics have led to applications as sneaker uppers and lining, outdoor sports gloves, dress shoes and tactical boots (Formula4 Media, 2016; Kuraray Co, inpress). Fig. 3.11 shows the cross-section of Clarino microfiber leather and its applications.

3.3.3 Reducing waste and environmental hazards

Minimizing waste and reducing environmental hazards during manufacturing processes and after producing the products are also important components of maintaining a balance between doing business and protecting the environment. A typical example is the Nike Flyknit and Flyprint technologies. They use computer-controlled knitting and 3D printing to strategically place polyester yarns or TPU filaments when forming the entire shoe upper. These technologies significantly help to reduce fabric waste which would otherwise be produced during cut and sew processes. Apart from Nike, there are other companies that are developing innovative clean technologies for textiles and garments.

Tiong Liong Industrial Co., Ltd. in Taichung City, Taiwan, has invented a new class of performance material for footwear, apparel and other technical gear called Ariaprene (Formula4 Media, 2016). Ariaprene is based on clean technology, in which the quick drying foam-core is toxic-free and used along with the bluesign certified solvent free lamination so that workers are not exposed to environmental hazards during the product manufacturing process (Tiong Liong Industrial, 2016). This technology is designed to feel and behave like a second skin for unrestricted movement, and is lightweight and highly flexible. Since it is eco-friendly and does not irritate the skin, Ariaprene is a hypoallergenic market solution for next-to-skin applications. In the activewear market, Ariaprene has been used for innovative applications. For example, Nike uses Ariaprene in its athletic training gear for lightweight and breathable paddings. Athleta is the first to use Ariaprene in the bra top of their QuickDri swim line because the hydrophobic and nontoxic foam is ideal for close-to skin applications.

Figure 3.12 Applications of Ariaprene. (A) Dakine Hellion knee pads: maintain the position of protective pad by the use of light weight, 4-way stretch Ariaprene, (B) Athleta swimwear: composite hydrophobic and non-toxic AriapreneÒ for close-to skin applications, and (C) Nike LunarGlide 6: Full Ariaprene upper with an engineered perforation pattern.
Source: Modified from Formula4 Media (2016).

DAKINE uses Ariaprene in their Slayer and Hellion knee pads for mountain biking as it provides the flexibility to support movement and allows breathability (Formula4 Media, 2016). Fig. 3.12 shows the products made of Ariaprene.

In addition, Bolger & O' Hearn, Inc., one of the most respected developers of substitutes for textile chemistries since the company is certified to follow bluesign regulations, has developed an advanced perfluorinated compound (PFC)-free water repellent finish called Altopel F^3 (the F^3 stands for Fluorine-Free Finish), which is both environmentally-friendly and long-lasting. Fluorinated materials are important in the textile industry to coat materials. The application of fluorinated materials may however require the dispersion of fluorinated surfactants. These surfactants are not biodegradable, and when they are discharged into groundwater or drinking water, both animals and humans can ingest the toxic chemicals which accumulate in their body. The toxicity of fluorinated materials has caused much concern around their implications on health and the environment. Therefore, the PFC-free water repellent finishing method is now being adopted by many in the textile industry since outdoor garments and activewear are usually water repellent. Besides, it is a durable water repellent (DWR) finish. When used on 100% nylon or 100% polyester, Altopel F^3 can last through 50 or more cycles of washings. On cotton, Altopel F^3 lasts through 20–30 washings. Also, Altopel F^3 does not affect the hand feel or breathability of the fabric (Bolger & O' Hearn, 2019).

Nau International Inc. has also launched PFC-free DWRs. They used a variety of PFC-free DWR alternatives such as bio-based and hydrocarbon polymer-based chemistries to provide a finishing with the highest level of water repellency but the lowest impact on the environment. Their PFC-free chemistries are assessed by accredited third party organizations such as bluesign, ZDHC, Standard 100 by Oeko-Tex, and globally recognized Restricted Substance Lists (Nau International Inc, 2016).

As for an eco-friendly finishing method that inhibits the bacteria that causes odors, Polygiene AB offers an innovative odor control technology called the Polygiene Odor Control Technology Wear More. Wash Less. Sweat itself is odorless but when it is trapped in the fibers of textiles, bacteria can grow in the warm and moist environment which results in a bad smell. They claimed that this odor forces 30% of people to discard their clothes prematurely (Polygiene, 2019). Polygiene AB uses naturally occurring silver salt (silver chloride) derived from recycled silver to provide odor control and inhibit bacterial growth. The silver salt technology is only active on the textiles themselves and is not nanosilver (Polygiene, 2019). This technology reduces the time, energy, and resources required to wash garments. They claim that washing one load less of laundry each week is equal to having 3 days of free time and saves up to 8000 L of water after 1 year. Also, less laundering can prolong the life of garments and reduce textile waste if the garments are not discarded due to fading, damage or stretching because of frequent washing. Wear More. Wash Less can be applied to a wide range of textiles for an active lifestyle including active and travel apparel, casual and daily wear garments, protective gear and gloves, and footwear. Wear More. Wash Less is also bluesign and OEKO-TEX approved, and garments treated with this Polygiene Odor Control Technology are recyclable (Formula4 Media, 2016).

3.4 Conclusion

The latest developments in fibers, yarns, fabrics, finishing processes and sustainable technologies that are being used to manufacture activewear have been discussed in this chapter. Useful information is provided which may inspire manufacturers, product developers, and designers to develop more innovative applications with new materials and technologies. The products and technologies elaborated here synthesize various functions which maximizes the benefits of activewear. For example, Spectra fiber is light in weight, high in strength and durable, which not only provides sensorial comfort but also protective functions. VORTEX yarn reduces pilling and is less hairy, thus providing a cotton-like texture which results in improved comfort and aesthetics. Apart from improved textile performance, some eco-technologies such as PFC-free water repellent finishings and innovative odor control technologies have reduced or no impacts on the environment. Unifi offers a series of recycled yarn from plastic bottles and Nike uses their own technologies to reduce industrial waste. These companies therefore inspire new perspectives in the development of activewear.

References

Barker, R. L. (2002). From fabric hand to thermal comfort: the evolving role of objective measurements in explaining human comfort response to textiles. *International Journal of Clothing Science and Technology*, *14*(3/4), 181–200.

Bolger & O' Hearn, Inc. (2019). Advanced FluorineFree WATER REPELLENT. Available from: https://bolgerohearn.com/f3-altopel/, Accessed 29 May 2019.

Bonaldi, R. R., Pak, P. H., De Lacerda, F. P., & Costa, N., Rhodia Poliamida E Especialidades SA (2018). Polyamide fibre with improved comfort management, process thereof and article made therefrom. U.S. Patent Application 15/759,446.

Bonser, R. H., & Dawson, C. (1999). The structural mechanical properties of down feathers and biomimicking natural insulation materials. *Journal of Materials Science Letters*, *18*(21), 1769–1770.

Cai, Q. X., Tian, F. Y.,& Xie, J. Y., G-FUN INDUSTRIAL Corp (2018). Process of manufacturing rain waterproof breathable fabric. US Patent 9,920,471.

Chandrasekaran, K., & Senthilkumar, M. (2018). Effect of herbal extract treatment on the moisture management properties of cotton knitted fabrics. *Journal of Natural Fibres*, 1–16.

Chen, Y., Yan, Y., Wu, B., & Tao, J. (2015, December). Summer sportswear fabrics research of jade fibre and comprehensive evaluation. In *2015 8th international symposium on computational intelligence and design (ISCID)* (Vol. 1, pp. 81–84). IEEE.

Classen, E. (2018). Comfort testing of textiles. In *Advanced characterization and testing of textiles* (pp. 59–69). Duxford, England: Woodhead Publishing.

Cotton Incorporated (2019). TransDRY® Technology. Available from: https://www.cottonworks.com/topics/fabric-technology/performance-technologies/transdry-technology/, Accessed 21 December 2019.

Cotton Incorporated (2019). STORM COTTON™ Technology. Available from: https://www.cottonworks.com/topics/fabric-technology/performance-technologies/storm-cotton/, Accessed 21 December 2019.

Cotton Incorporated (2019). WICKING WINDOWS (Technology. Available from: https://www.cottonworks.com/topics/fabric-technology/performance-technologies/wicking-windows-technology/, Accessed 21 December 2019.

Cui, S. L., Zhu, J. P., Li, B. M., Wang, S. P., & Liu, Y. (2012). Structure and physical properties of jade fibre. In *Advanced Materials Research* (pp. 767–771). (441). Trans Tech Publications.

Daniels, J. (2018). Flyknit, the new technology of your trainers. Available from: https://www.luxurylifestylemag.co.uk/style-and-beauty/flyknit-the-new-technology-of-your-trainers/, Accessed 23 May 2019.

DSM (2016). Dyneema-reinforced full outfit gives increased crash protection for Giant Alpecin at Tour de France 2016. Available from: https://www.dsm.com/products/dyneema/en_GB/about/press-releases/press-releases-2016/dyneema-reinforced-full-outfit-gives-increased-crash-protection.html Accessed 23 May 2019.

DSM. Dyneema Fibre. Available from: https://www.dsm.com/products/dyneema/en_GB/technologies/dyneema-form-factors/fibre.html Accessed 23 May 2019.

Fibre-line (2019). UHMWPE. Available from: https://www.fibre-line.com/en/fibres/uhmwpe Accessed 23 May 2019.

Formula4 Media, LLC. (2016).Textile insight®. Available from: http://c919297.r97.cf2.rackcdn.com/8p7j1vxigmue54r2wsdmqrzagoh9i9-optimized-pub.pdf, Accessed 29 May 2019.

Gonzalez, R. (2017). Patagonia Perfects the Puffy Jacket with Ultra-Light Insulation. Available from: https://www.wired.com/story/patagonia-perfects-the-puffy-jacket/, Accessed 17 May 2019.

Grossman, M. (2017). What is Neoprene? From Scuba Diving to Avant Garde. Available from: https://www.contrado.co.uk/blog/what-is-neoprene/, Accessed 6 May 2019.

Harlin, A., Jussila, K., & Ilen, E. (2019). Sports textiles and comfort aspects. In *High Performance Technical Textiles* (pp. 37–67). Hoboken, NJ: Wiley.

Harvest Consumer Insulation, LLC (2015a). Durable Warmth. Available from: http://climashield.com/durable-warmth, Accessed 17 May 2019.

Harvest Consumer Insulation, LLC (2015b). Continuous Filament. Available from: http://climashield.com/continuous-filament, Accessed 17 May 2019.

Havenith, G., Bröde, P., Hartog, E. D., Kuklane, K., Holmer, I., Rossi, R. M., Richards, M., Farnworth, B., & Wang, X. (2013). Evaporative cooling: Effective latent heat of evaporation in relation to evaporation distance from the skin. *Journal of Applied Physiology*, *114*(6), 778–785.

Hayes, S. G., & Venkatraman, P. (Eds.). (2016). *Materials and technology for sportswear and performance apparel*. Boca Raton, FL: CRC Press.

Honeywell International Inc. (2018). Science behind spectra fibre. Available from: https://www.spectra-fibre.com/science-behind-spectra-fibre/ Accessed 6 May 2019.

HYOSUNG TNC (2018). cotna. Available from: http://www.hyosungtnc.com/en/fibre/polyester.do, Accessed 17 May 2019.

Imperatives, S. (1987). Report of the World Commission on Environment and Development: Our common future.

Jenkins, E. (2018). High-performance apparel for outdoor pursuits. In *High-Performance Apparel* (pp. 421–449). Duxford, England: Woodhead Publishing.

Jones, R. (2013). KNOW YOUR TECH: Nike Flyknit. Available from: https://www.complex.com/sneakers/2013/07/know-your-tech-nike-flyknit, Accessed 23 May 2019.

Jun, Y., Park, C. H., Shim, H., & Kang, T. J. (2009). Thermal comfort properties of wearing caps from various textiles. *Textile Research Journal*, *79*(2), 179–189.

Kanjana, S., & Nalankilli, G. (2018). Smart, waterproof, breathable sportswear—A review. *Journal of Textile & Apparel Technology & Management (JTATM)*, *10*(3), 591–600.

Ko, E., Taylor, C. R., Sung, H., Lee, J., Wagner, U., Navarro, D. M. C., & Wang, F. (2012). Global marketing segmentation usefulness in the sportswear industry. *Journal of Business Research*, *65*(11), 1565–1575.

Krishnaveni, K. (2017). Effect of imparting antimicrobial coating on organic cotton fabric using yashtimadhu for medical application. *Journal of Textile Engineering & Fashion Technology*, *3*(3), 650–652.

Kuraray Co., Ltd. TIRRENINA™ environmentally friendly microfibre suede. Available from: https://www.kuraray.eu/en/produkte/product-ranges/tirreninatm/, Accessed 29 May 2019.

Matusiak, M. (2010). Thermal comfort index as a method of assessing the thermal comfort of textile materials. *Fibres & Textiles in Eastern Europe*, *18*(2), 79.

Mondal, M. I. H., & Saha, J. (2019). Antimicrobial, UV resistant and thermal comfort properties of chitosan-and aloe vera-modified cotton woven fabric. *Journal of Polymers and the Environment*, *27*(2), 405–420.

Murata Machinery, Ltd. (2014). Football brand "Umbro" sportswear now available. Available from: http://www.muratec-vortex.com/information/659/, Accessed 17 May 2019.

Murata Machinery, Ltd. What is VORTEX yarn. Available from: http://www.muratec-vortex.com/vortex01.html, Accessed 17 May 2019.

Murata Machinery, Ltd. Features of VORTEX yarns. Available from: http://www.muratec-vortex.com/feature06.html, Accessed 17 May 2019.

Nau International, Inc. (2016). 37.5 Technology. Available from: https://www.nau.com/our-fabrics/synthetic-insulation, Accessed 6 May 2019.

Nau International, Inc. (2016). Recycled materials. Available from: https://www.nau.com/our-fabrics/recycled-downs, Accessed 29 May 2019.

Nau International, Inc. (2016). PFC-FREE DWR. Available from: https://www.nau.com/our-fabrics/pfc-free-dwr, Accessed 29 May 2019.

Neshin Spinning Co. Ltd. Cooling Yarn. Available from: http://www.neshin.com.tw/en/product-c76817/Cooling-Yarn.html Accessed 6 May 2019.

Nike, Inc. (2013). How Nike Flyknit revolutionized the age-old craft of shoemaking. Available from: https://www.theguardian.com/sustainable-business/partner-zone-nike1, Accessed 23 May 2019.

Nike, Inc. (2014). Nike Football Accelerates Innovation with 3D printed "Concept Cleat" for Shuttle. Available from: https://news.nike.com/news/nike-football-accelerates-innovation-with-3d-printed-concept-cleat-for-shuttle, Accessed 29 May 2019.

Nike, Inc. (2018b). Nike Flyprint is the First Performance 3D Printed Textile Upper. Available from: https://news.nike.com/news/nike-flyprint-3d-printed-textile, Accessed 29 May 2019.

NIKE, Inc. (2019). Celebrating the Sports Bra. Available from: https://news.nike.com/news/nike-flyknit-sports-bra, Accessed 23 May 2019.

NIKE, Inc. (2013). Nike debuts first-ever football cleat built using 3D printing technology. Available from: https://news.nike.com/news/nike-debuts-first-ever-football-cleat-built-using-3d-printing-technology, Accessed 29 May 2019.

O'Sullivan, G. A., Hanlon, C., Spaaij, R., & Westerbeek, H. (2017). Women's activewear trends and drivers: A systematic review. *Journal of Fashion Marketing and Management: An International Journal*, *21*(1), 2–15.

Parsons, K. (2014). *Human thermal environments: The effects of hot, moderate, and cold environments on human health, comfort, and performance*. CRC Press.

Patagonia (2018). It All Adds Up to Nothing: Forging the Micro Puff. Available from: https://www.patagonia.com/blog/2018/02/it-all-adds-up-to-nothing-forging-the-micro-puff/, Accessed 17 May 2019.

Patagonia. PrimaLoft® Insulation. Available from: https://www.patagonia.com/primaloft.html, Accessed 17 May 2019.

Performance Sport Limited. COOLJADE FABRIC™ (2019). Available from: https://psport.com.hk/pages/virus-techology Accessed 6 May 2019.

Polygiene AB. (2019). Polygiene® How it works. PFC-FREE DWR. Available from: https://polygiene.com/how-it-works/, Accessed 1 June 2019.

PrimaLoft, Inc. (2018). Primaloft® Insulation the World's Best Down Alternative. Available from: http://www.primaloft.com/insulation, Accessed 17 May 2019.

Qingdao Bettertex Fibre Co., Ltd. Icejade. Available from: http://www.bettertex.cn/en/product/brand/34 Accessed 6 May 2019.

Raccuglia, M., Sales, B., Heyde, C., Havenith, G., & Hodder, S. (2018). Clothing comfort during physical exercise–determining the critical factors. *Applied Ergonomics*, *73*, 33–41.

Shishoo, R. (2015). Introduction to textiles in sport. In *Textiles for Sportswear* (pp. 3–16). Cambridge, England: Elsevier.

The LYCRA Company. THE COOLMAX® BRAND. Available from: http://coolmax.com/en/About-the-COOLMAX-Brand/The-COOLMAX-brand, Accessed 17 May 2019.

Tiong Liong Industrial Co., Ltd. (2016). Ariaprene® technology. Available from: http://www.ariaprene.com/index.php/technology/, Accessed 29 May 2019.

Ünal, Z. B., & Eren, E. R. (2018). The use of neoprene fabric evaluation in terms of comfort in child tracksuit production. *Industria Textila*, *69*(5), 400–405.

Unifi, Inc. (2019a). Repreve®/ How it's made. Available from: https://unifi.com/repreve/how-its-made, Accessed 29 May 2019.

Unifi, Inc. (2019b). Recycled materials. Available from: https://unifi.com/products/recycled-materials, Accessed 29 May 2019.

Uren, N., & Okur, A. (2019). Analysis and improvement of tactile comfort and low-stress mechanical properties of denim fabrics. *Textile Research Journal*, *89*(23–24), 4842–4857.

Wang, W., Yao, L., Cheng, C. Y., Zhang, T., Atsumi, H., Wang, L., Wang, G., Anilionyte, O., Steiner, H., Ou, J., & Zhou, K. (2017). Harnessing the hygroscopic and biofluorescent behaviors of genetically tractable microbial cells to design biohybrid wearables. *Science Advances*, *3*(5), p.e1601984.

Wang, X., Huang, Z., Miao, D., Zhao, J., Yu, J., & Ding, B. (2018). Biomimetic fibrous murray membranes with ultrafast water transport and evaporation for smart moisture-wicking fabrics. *ACS Nano*, *13*(2), 1060–1070.

Wang, X., Huang, Z., Miao, D., Zhao, J., Yu, J., & Ding, B. (2019). Biomimetic fibrous murray membranes with ultrafast water transport and evaporation for smart moisture-wicking fabrics. *ACS Nano*, *13*(2), 1060–1070.

Yao, L., Wang, W., Wang, G., Steiner, H., Cheng, C. Y., Ou, J., Anilionyte, O., Ishii, H. (2015). Biologic. Available from: https://tangible.media.mit.edu/project/biologic/, Accessed 29 May 2019.

Yao, L., Ou, J., Wang, G., Cheng, C. Y., Wang, W., Steiner, H., & Ishii, H. (2015). bioPrint: A liquid deposition printing system for natural actuators. *3D Printing and Additive Manufacturing*, *2*(4), 168–179.

Yu, Z. C., Zhang, J. F., Lou, C. W., He, H. L., Chen, A. P., & Lin, J. H. (2015). Determination of electromagnetic shielding and antibacterial properties of multifunctional warp-knitted fabrics. *The Journal of the Textile Institute*, *106*(11), 1203–1211.

Zurcher Wray, A., & Nelson Hodges, N. (2008). Response to activewear apparel advertisements by US baby boomers: An examination of cognitive versus chronological age factors. *Journal of Fashion Marketing and Management: An International Journal*, *12*(1), 8–23.

Narrow fabric elastics and accessories for activewear

4

Joanne Yip, Olivia Ho-Yi Fung, Tsz-ling Ng
Institute of Textiles and Clothing, The Hong Kong Polytechnic University, Hung Hom, Hong Kong

4.1 Introduction

In recent years, interest in health and fitness, along with participation in sporting activities, has been increasing and is becoming more widespread among the general population. This trend has therefore caused a rapid expansion of the activewear market (McCann, 2005). Activewear plays an important role in optimizing the performance of athletes by maximizing their comfort, minimizing the risk of injury as well as muscle fatigue, reducing friction/drag, and providing freedom of movement (Voyce, Dafniotis, & Towlson, 2005). There have also been some interesting developments that are taking place in regard to intelligent textiles and interactive materials, which have a huge market potential in the activewear sector (Shishoo, 2005). In this chapter, developments in narrow fabric elastics and accessories for activewear are discussed with examples based on three main factors: (1) functionality (i.e., including support, ease of movement, and user-friendliness), (2) esthetics, and (3) smart technology.

4.2 Functionality in activewear

Athletes and sports enthusiasts are constantly demanding for better performing activewear even though their current garments may still be serviceable, especially sports bras. Improvements in the functionalities of activewear can therefore greatly meet user expectations and in terms of sports bras, better breast support, ease of movement, and user-friendliness are usually the primary requests. In this section, the different developments in elastics and accessories with respect to the mentioned functions are discussed as follows.

4.2.1 Support of the breasts

Vertical breast displacement during physical activity has been associated with exercise-induced breast discomfort. In other words, larger vertical movement of the breasts result in more breast discomfort (McGhee, Steele, Zealey, & Takacs, 2013). Breast support is therefore one of the most important design considerations for sports bra, particularly for women with large breasts. According to McGhee and Steele (2010), insufficient breast support can give rise to the development of musculoskeletal pain

Latest Material and Technological Developments for Activewear. http://dx.doi.org/10.1016/B978-0-12-819492-8.00004-1

and limit the capability of women to engage in physical activity. In comparison to daily and fashion bras, sports bras excel more at reducing vertical breast displacement and, in turn, enhancing comfort during exercise (McGhee et al., 2013).

The bra straps are an important feature of a sports bra, and only secondary to the bra band in providing support to the breasts. Conventionally, the straps of sports bra use the same stretch fabric as the body of the bra or elastics with slider for length adjustment. The prior has a fixed strap length and cannot be adjusted to fit the wearer properly. Hence, the straps dig into the shoulders and create discomfort if they are too tight, and slip off the shoulders if they are too loose. The length of the straps can be adjusted for the latter design; however, the slider may at times slides along the strap and therefore fails to hold the strap length in position during intensive physical movement in sporting activities. To address these shortcomings, a variety of different developments have emerged, which will be subsequently elaborated.

4.2.1.1 Slip resistance slider

Fildan Accessories Corporation has developed a device called the Flex'n Grip, which comprises a slider (Fig. 4.1) and a link for the shoulder strap of a sports bra (Fig. 4.2). The Flex'n Grip is curved in shape with grooves along the left and right edges that grip the straps and prevent them from sliding. The curved design also prevents the edges of the slider/link from digging into the skin, which is a problem of conventional flat-shaped sliders (Fig. 4.3).

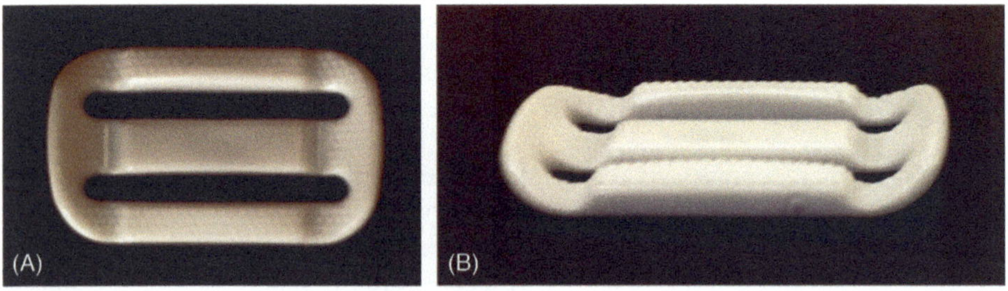

Figure 4.1 Flex'n Grip: slider (A) top view and (B) side view.

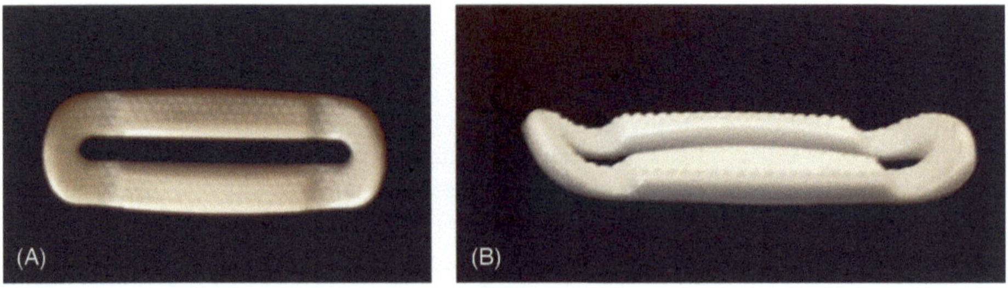

Figure 4.2 Flex'n Grip: link (A) top view and (B) side view.

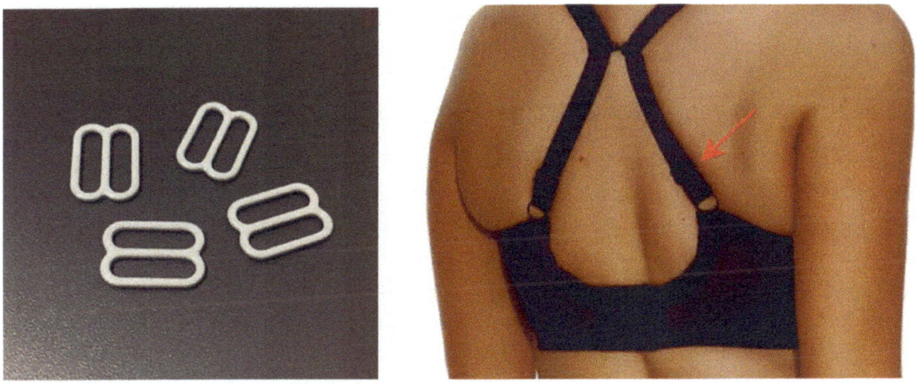

Figure 4.3 Conventional flat-shaped slider: digs into skin.

4.2.1.2 Welded adjustable strap with loops/buttons

Fildan Accessories Corporation has also developed the LOOP-TAPE adjuster strap; see Fig. 4.4, which is a strap that has several loops/buttons welded onto its surface with 9-hooks/buttons on one end which are ultrasonically welded, so that no stitching is required. The 9-hooks/buttons on the end of the strap are designed to attach onto one of the loops/buttons along the strap so that a suitable strap length is possible. This design can address the problem of using a conventional slider which can slide down the

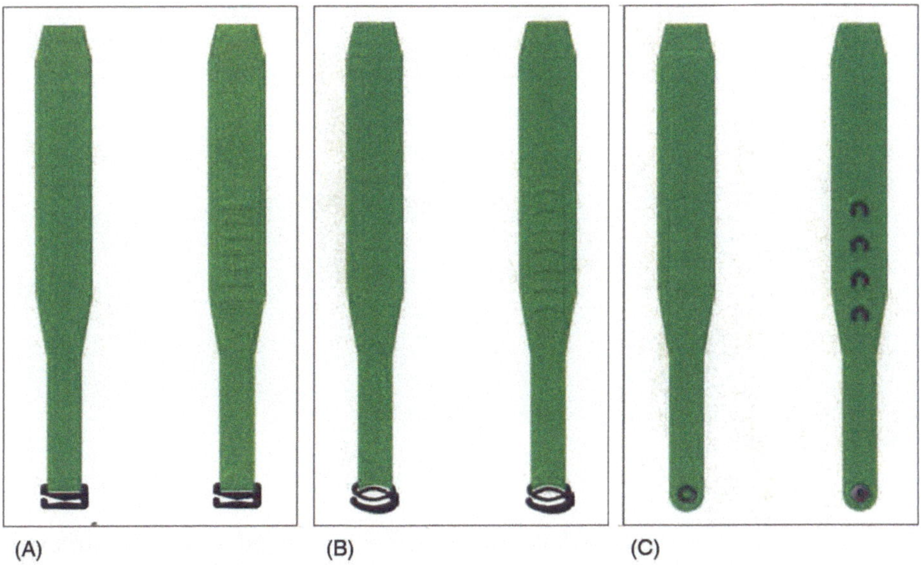

Figure 4.4 LOOP-TAPE adjuster strap: (A) straight loops, (B) curved loops, and (C) 4 HEY buttons.

strap. Furthermore, the LOOP-TAPE adjuster strap is cushioned with a 2 mm spacer for better comfort and less pressure on the shoulders.

4.2.1.3 Woven adjustable strap/band with loops

Pioneer Elastic Ltd. has developed an adjustable elastic band; see Fig. 4.5 (Liu & Xie, 2010) with a similar design concept as that of the LOOP-TAPE adjuster strap. The only difference is that the band is woven instead of ultrasonically welded. The woven band comprises two layers. The bottom layer is the base to form loops on the surface of the top layer. As the bands are knitted one after one successively, the larger loops have to be trimmed to separate them individually. A 9-hook latch is sewn onto the trimmed end and fixed onto one of the smaller loops based on the desired length, which can be adjusted later based on need. In comparison to the LOOP-TAPE adjuster strap, the production lead time of the Pioneer Elastic Ltd. adjustable elastic band is shorter because the entire band can be completed with one weaving process. The LOOP-TAPE adjuster strap requires an additional step of ultrasonically welding

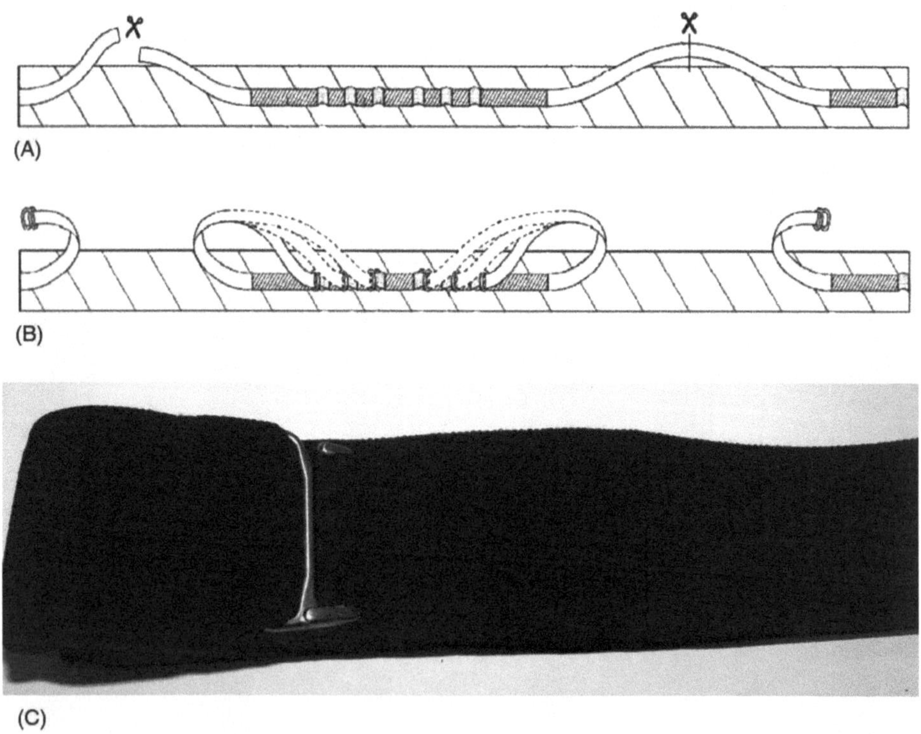

Figure 4.5 Adjustable elastic band: (A) larger loops trimmed, and bottom layer trimmed from middle (i.e., to separate the bands individually), (B) end of top layer fixed onto one of the smaller loops with 9-hook in accordance with desired length, and (C) a sample of adjustable elastic band.

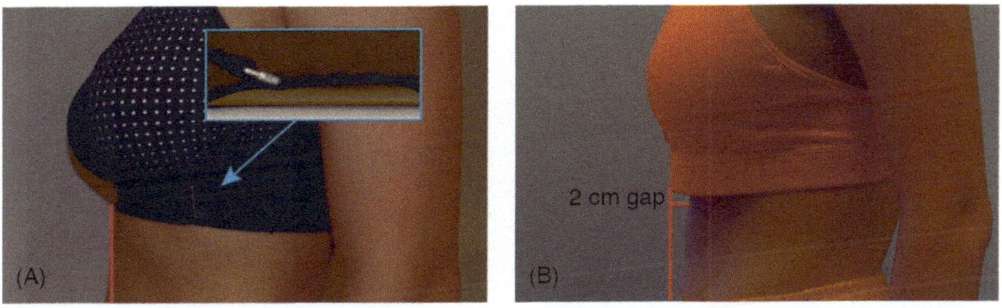

Figure 4.6 Underband of sports bra: (A) use of adjustable elastic bra band and (B) conventional design which does not allow for length adjustments.

the loops/buttons. Furthermore, patterns can be incorporated into the adjustable elastic band through jacquard weaving for esthetics.

The application of this design by Pioneer Elastic Ltd. is not limited to shoulder straps, but also the underband of sports bra as well as the waistband of track pants or tights. Fig. 4.6 shows the difference in fit between two sports bras; Fig. 4.6A shows the sports bra with the adjustable elastic bra band and Fig. 4.6B shows the sports bra without the adjustable elastic bra band. The latter shows a gap of 2 cm between the underband and torso, but not the former. Since sports bras usually do not offer a large variety of sizes as is the case for daily or fashion bras, that is, sizes S, M, and L in comparison to 32B, 32C, etc., an adjustable elastic bra band can remedy the ill-fitting problem that arises from the underband and provides better support to the breasts.

4.2.2 Ease of movement

To prevent the straps of a sports bra from slipping off the shoulders during intensive movement in physical activities and sports, women are often inclined to tighten them. However, this would affect the bra fit. The underband might lift and therefore would not sit at the same level as the inframammary fold. As a result, this would reduce the support given to the breasts. Also, the excessive pressure that is exerted through the bra straps onto the shoulders can create deep grooves on them, and cause soreness over long durations of bra wear. In view of this, some commercially available sports bras have introduced shoulder straps that incorporate silicone gel pads as a form of cushioning at the rear surface. This intends to address the slipping-off problem of the shoulder straps by increasing friction at the bra strap–shoulder interface (Coltman, McGhee, & Steele, 2015), so that sports and exercise can be performed with ease. Unfortunately, women who are allergic to silicone cannot use this product, and silicone cracks after several uses.

4.2.2.1 Orientation of shoulder straps

As shown in Fig. 4.7, the most common way of preventing the shoulder straps of sports bras from slipping off the shoulders is to orient them diagonally toward the center of the back rather than vertically. The design not only helps to preventing the

Figure 4.7 Sports bra with different orientations of shoulder straps for preventing straps from slipping off the shoulders.

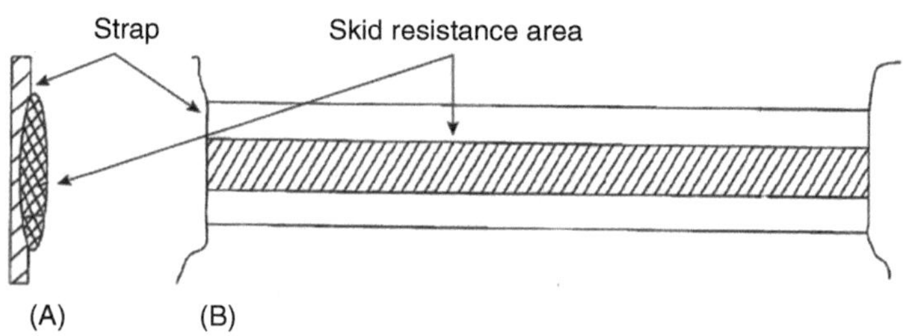

Figure 4.8 Slip resistance strap: (A) side view and (B) bird's eye view.

straps from slipping off the shoulders, but also gives fashionable appearance to the wearers.

4.2.2.2 Slip resistance strap

In considering the shortcomings of the silicone gel pads, Pioneer Elastic Ltd. has developed a slip resistance strap (Liu, 2009) (Fig. 4.8). The skid resistance area on the strap is protruded, knitted with spandex, and formed at the same time when the strap is knitted. Moreover, the protrusions can have different patterns, namely dots, stripes, hearts, etc. for esthetics. Unlike silicone gel pads, no additional manufacturing process is required to create the slip resistance effect, and thus the production lead time is shorter.

4.2.3 User-friendliness

Bra fasteners are usually rows of small-sized hooks and eyes, which are located at the back of the bra. To some people, these kinds of fasteners are not the most user-friendly design because the act of fastening and unfastening the bra at the back is not an easy task. Hence, various developments have been subsequently introduced to address this issue, which are elaborated as follows.

4.2.3.1 Special hook and eye designs

Fildan Accessories Corporation has developed a new type of hook and eye which they call the OFF-Hook and Eye (HEY) closing system (Fildan & Wanzenböck, 2017) (Fig. 4.9). Since the design of the OFF-HEY closing system only has one long hook instead of many little hooks, the wearer can fasten and unfasten her bra with ease and the design is thus more user-friendly. The frame of the hooks and eyes is mounted directly onto a fabric strip which can be either cushioned or noncushioned depending on the need. The thickness is around 4–5 mm when the OFF-HEY is fastened, which is therefore thin, has a low profile and not visible under most garments. Furthermore, the OFF-HEY closing system can be used in sports bras because it is capable of withstanding high tensile force, that is, a maximum of 30 lbs/13.6 kg, which can result from the extensive range of various movements during sporting activities and exercise.

Fildan Accessories Corporation subsequently developed another type of hook and eye closing system which they called TOO-HEY (Fildan & Wanzenböck, 2018) (Fig. 4.10). The design concept is similar to the OFF-HEY closing system in that there is only one hook instead of many hooks for ease of fastening. However, this system can

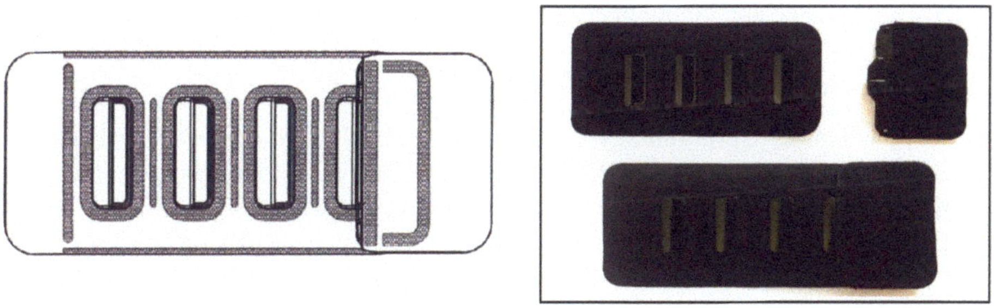

Figure 4.9 OFF-HEY closing system.

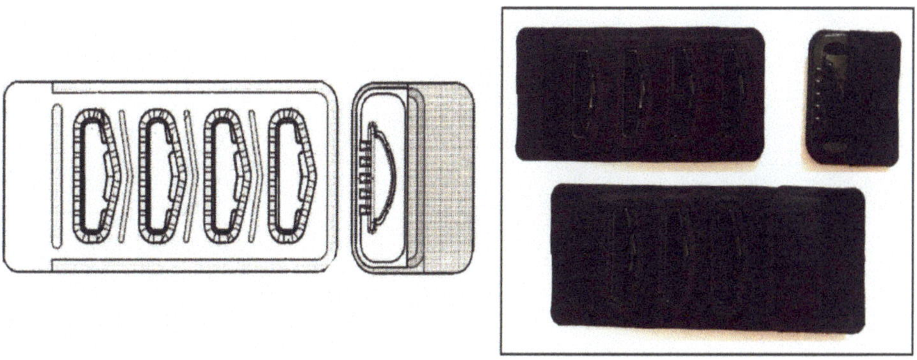

Figure 4.10 TOO-HEY closing system.

only withstand a maximum tensile force at of 25 lbs/11.4 kg instead of 30 lbs/13.6 kg. Moreover, all of the TOO-HEY components are made of polyamide (PA) and thus recyclable. To attach the eyes onto the frame more firmly on the fabric strip, they are welded together with grooves. Ribs are formed at the edges of the hook in order to reduce the tendency of the contraption to bend when force is applied. This can also enable the hook fastener to be produced with a slimmer profile without a reduction in its strength. The hook and eye fasteners are chevron shaped to help guide the tip of the hook into the center of the eye, which further facilitates the ease of fastening.

4.2.3.2 *Hidden front closure design*

Users may prefer bras with front closures because they are placed at the front instead of the back of the bra which makes fastening an easier task. However, bras with front closures currently available in the market are usually visible, which may be embarrassing and undesirable for some wearers. Fildan Accessories Corporation has once again addressed this dilemma by offering a hidden front closure, which are two closure halves that can be almost undetectable when closed together. The components are hidden at the back of the front fabric pieces and only a seam line is shown at the center of the bra gore (Fildan & Wanzenböck, 2004) (Fig. 4.11). There are two ways of attaching the two closure halves onto the gore fabric: one is by stitching and the other is by welding. As stitches will show through the fabric, welding can provide a cleaner appearance and completely hide the front closure.

4.2.3.3 *Closure with magnetic design*

Fildan Accessories Corporation has also developed a super magnetic back closure for ease of fastening (Fildan & Wanzenböck, 2012) (Fig. 4.12). When the male and female magnets are close to each other, they automatically engage. The super magnetic closure is designed to be locked in three directions under tension from wear. The

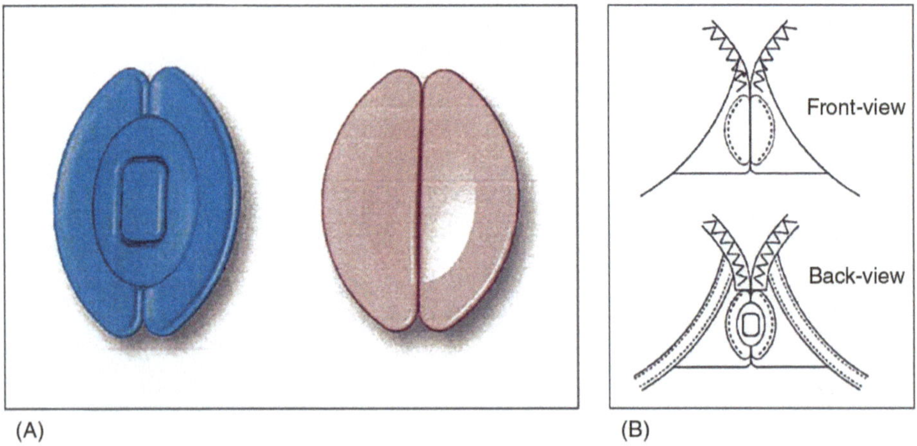

Figure 4.11 (A) Hidden front closure and (B) attaching two closure halves as hidden front closure of bra by stitching.

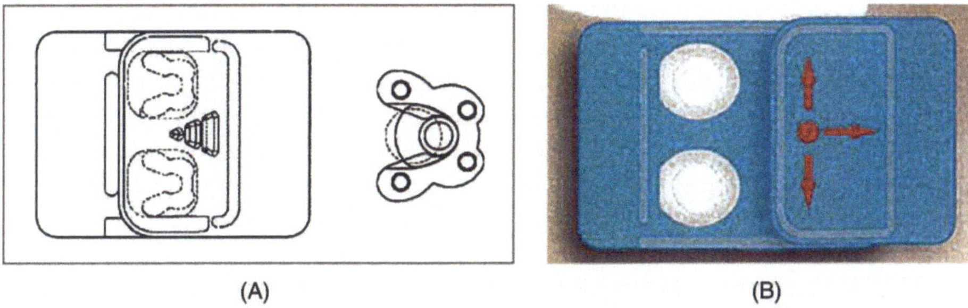

Figure 4.12 Super magnetic closure: (A) construction and (B) illustration on how the closure secures in three axes under tension from wear.

maximum tensile force that the closure can withstand is 25 lbs/11.3 kg. The magnet used is also waterproof and can therefore be used in swimwear.

4.3 Esthetics

Activewear with only functionality and comfort cannot entirely satisfy the needs of the end-users, as appearance is also an important element. Nonetheless, activewear design would be considered a failure if for example, a snowboarder looks "cool" in it, but feels uncomfortable and frustrated because the garment does not have a good performance. Therefore, it is essential to balance technical attributes and esthetics (McCann, 2005). The following will elaborate on some of the recent esthetic developments in elastic straps.

4.3.1 Creating cut-out patterns on elastic straps with no deformation

Cut-out patterns can be easily created on a rigid woven strap which has no elasticity. However, if they are created on an elastic strap, the yarn of the strap which is elastic will spring back and thus the pattern will deform. In view of this, Pioneer Elastic Ltd. has developed a method to create cut-out patterns on elastic straps with no deformation (Fig. 4.13) (Lin & Zou, 2007). First, the elastic strap is fabricated from heat-melt

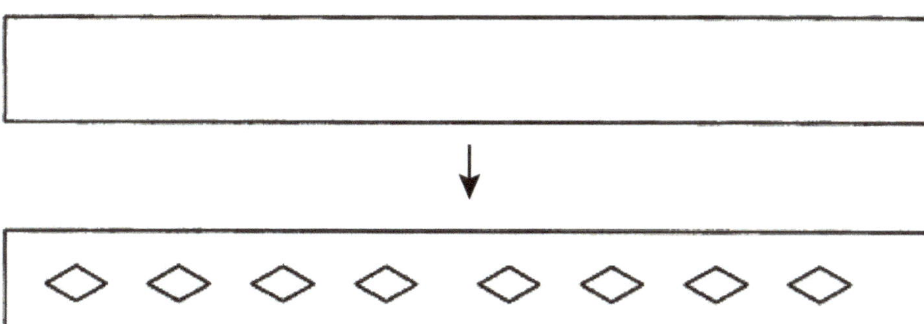

Figure 4.13 Cut-out pattern on elastic strap.

and elastic yarns. The heat-melt yarn can be made from any kind of human-made fiber, and the elastic yarn can either be Spandex, lycra, or elastane. Next, the elastic strap has to be heat-set, so that the heat-melt yarns melt and adhere to the elastic yarns nearby. This process can help to prevent the elastic yarns from springing back and reducing the rate of shrinkage of the entire elastic strap. Finally, patterns are ultrasonically cut from the elastic strap. The cut-out pattern has cleanly trimmed edges with no frays and deformation. Also, the elasticity of the strap is maintained.

4.3.2 Elastic strap pattern changes based on viewing angle

Color changes are the most common means of providing different visual effects in current commercially available textile products. In general, this can be done by the application of color changing powder, light sensitive powder or heat sensitive material to the yarn or textile. However, the downside of this technique is that it has to be applied to the entire rather than part of the fabric. The fabrics are usually woven with no elasticity and, also, the production of which is complicated. Additionally, some other textile products integrate electric circuits for a color changing effect, but these are not comfortable to wear and the production cost is high.

Pioneer Elastic Ltd. has thus developed an elastic strap with changes in patterns and colors based on the viewing angle (Xie, 2016) (Fig. 4.14). The strap is composed

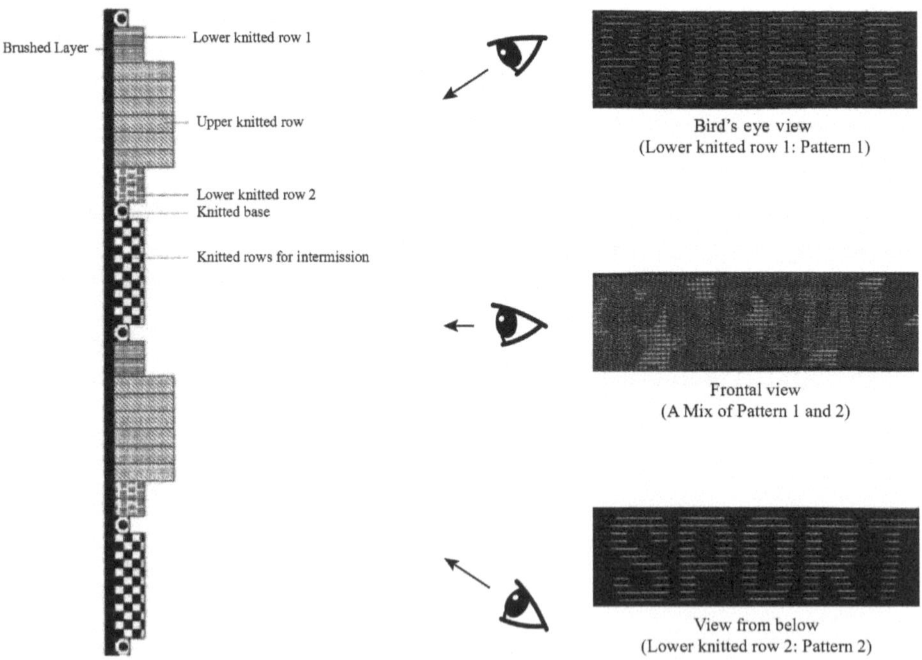

Figure 4.14 Elastic strap with changing patterns and colors based on viewing angle: cross-section of width of strap (left), and examples of patterns and colors that appear on strap from three different viewing angles (right).

of upper knitted rows, lower knitted rows 1, lower knitted rows 2, intermittent knitted rows (optional), a knitted base and a brushed layer. The lower knitted rows 1 are to form Pattern 1, and the lower knitted rows 2 for Pattern 2. They are both knitted on and closely against the knitted base. The upper knitted rows are also knitted on the knitted base but higher than the lower knitted rows, hence forming an arch. The purpose of the upper knitted rows is to block the view of the lower knitted rows from different angles. As shown in Fig. 4.14, the lower knitted rows 1 are consistently stitched at the top of each section of the upper knitted rows, whereas the lower knitted rows 2 are consistently stitched at the bottom of each section of the upper knitted rows. Therefore, if the elastic strap is viewed from a bird's eye view, only Pattern 1 which is formed with lower knitted rows 1 can be seen. Similarly, if the elastic strap is viewed from below, only Pattern 2 which is formed with lower knitted rows 2 can be seen. Intermittent rows can also be included but are optional. They have the same height as lower knitted rows 1 and 2 and are consistently stitched between them. Lastly, there is a brushed layer that covers the yarns on the wrong side of the knitted base which also serves to enhance comfort when the fabric comes into contact with the skin.

Fig. 4.15 shows the cross-section of the length of the strap to provide a visual aid on the construction of the strap. The black dots are the weft yarns of the knitted base. Arches are formed by the upper knitted row that is stitched at a height greater than the lower knitted row. The pattern formation is based on how the yarns of the lower knitted rows form along the knitted base. In other words, the color of the yarn will be showing when the yarn goes on top of the base, whereas the color of the base will be showing when the yarn goes beneath the base. In the case of Fig. 4.15, the yarn of the lower knitting row passes on top of 3 weft yarns, then beneath 1 weft yarn of the base, and keeps repeating. Since the yarn of the lower knitting row has a longer length on the top of the base than beneath, the strap will show mainly the color of the lower knitting row rather than the color of the knitting base. Besides, how the yarn of the knitted rows intertwines with the weft yarns of the knitted base is based on the design and left to imagination, that is, the yarn of the knitted rows can pass over the top of any number of weft yarns and beneath them.

It is also not necessary for the knitting rows to align parallel to the long edge of the strap like the above case (Figs. 4.14 and 4.15), since different orientations can generate different effects; see Fig. 4.16. Nonetheless, the knitted rows must parallel each other.

The recommended materials for fabricating the strap include stiff polyester yarn for the upper and lower knitted rows, nylon elastic yarn for the knitted base as well as the intermittent knitted rows, and nylon yarn for the brushed layer. Stiff yarn must

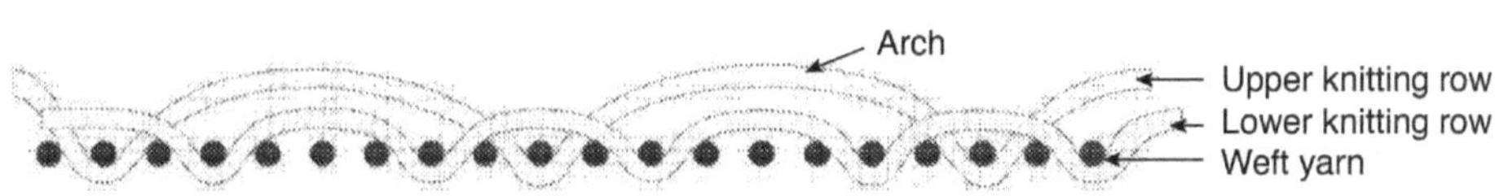

Figure 4.15 Elastic strap with changing patterns and colors based on viewing angle: cross-section of length of strap.

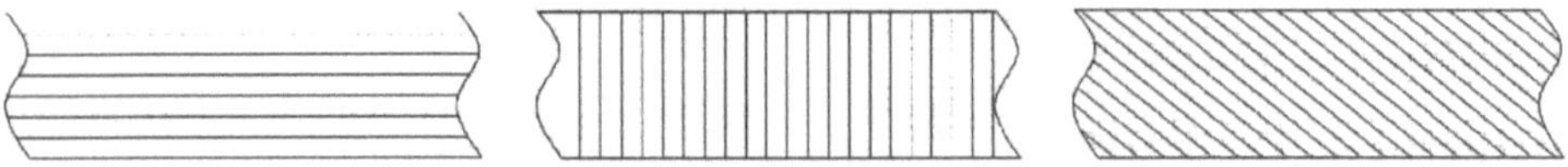

Figure 4.16 Elastic strap with changing patterns and colors based on viewing angle: knitted rows with different orientations for different effects.

be used for the upper and lower knitted rows to ensure that they can remain or revert back to their original shape after continuous washes and wear, so that the visual effect of the strap will not be lost.

This development of a visual effect on an elastic strap does not require heat/light sensitive materials or an additional process. Instead, the visual effect is created by the different height arrangements of the knitted rows, that is, yarn that is knitted into rows at different heights, and thus only a single knitting process is necessary to produce the entire strap. This method is therefore low in production cost. Moreover, the strap is comfortable to wear and the patterns and colors are limitless.

4.4 Smart technology

The physiological parameters of athletes, such as their rate of breathing, heart rate, galvanic skin response (GSR), etc. are conventionally measured by using electrodes that are coupled to them via leads. This is meant to monitor their health and wellness during exercise. However, electrodes can restrict the movement of the users and cause discomfort, and as a result, it is relatively challenging to monitor the real-time physiological reactions of athletes over extended periods of time. To address this issue, various textile-based electrode systems have been developed and integrated into performance activewear in recent years. The electrodes can be printed and laminated onto the textile, or stitched or sewn into the garment itself, and have to be in contact with the users (their skin) in order to function. This development also allows one or more physiological parameters of the users to be measured at the same time, as it is feasible to incorporate more than one electrode system into the garment. Nonetheless, wearable textile-based electrode systems still suffer from numerous shortcomings. First, stitched or sewn electrodes can chafe the skin and cause discomfort, especially during workouts and exercise due to intensive body movement. Furthermore, they are prone to wear and tear because of repeated use and washing; consequently, this can reduce the longevity of the electrode system. The overall production cost is also comparatively higher for stitched or sewn electrodes. Second, most textile-based electrode systems are configured to work with only two electrodes. In such a case, the electrodes generally need to be placed closely to each other, so that they can be coupled with an electrical device which is located between them in order to ensure that stable signals are produced. The electrical device can be a sensor or a processing module.

4.4.1 Knitted wearable textile-based electrode systems

In view of the need for textile-based electrode systems that minimize impacts on the perceived comfort of users, have more longevity, and provide a higher signal quality as well as accuracy, OMsignal Inc. has developed an improved textile-based electrode system (Berzowska, Chanay, Menard, & Nurkka, 2013) that can be constructed in three different ways. The first method uses three layers of fabric (Fig. 4.17). The first layer of fabric is configured with a knitted electrode in the inner surface which comes in contact with the user. On top of this layer is the second layer of fabric, which has a knitted conductive pathway that is electrically coupled to the electrode in the first layer of fabric. Then on top of these two layers is the third layer of fabric, which has a connector that is configured to be electrically coupled to the knitted conductive pathway in the second layer of fabric. The electrode in this system is an electrical conductor that is configured to come into contact with nonmetallic surfaces, that is, the user in this case, and measure the electrical signals that correspond to one or more physiological parameters of the user.

The second way of constructing this knitted wearable textile-based electrode system is that all of the components that form the system are knitted onto one single piece of

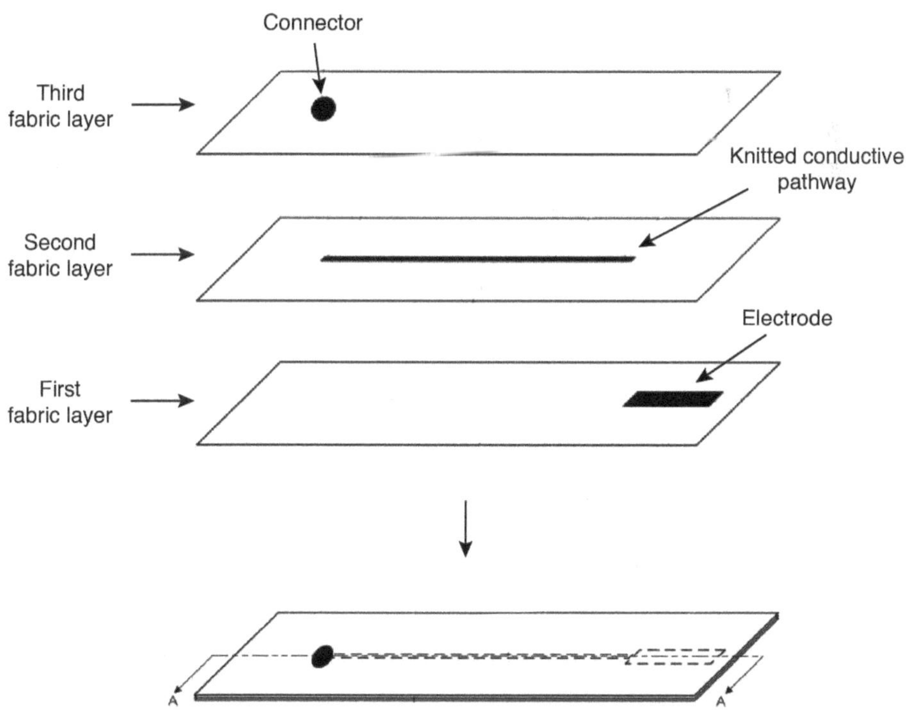

Figure 4.17 Knitted wearable textile-based electrode system: first method of construction with three separate layers of fabric.

fabric rather than using three separate layers of fabric (Fig. 4.18). There are three fabric portions on the fabric. The first fabric portion includes a knitted conductive pathway, the second fabric portion includes a knitted electrode, and the third fabric portion includes a connector. The process begins by folding the first fabric portion over the second fabric portion along the first fold line, so that the knitted conductive pathway on the first fabric portion is electrically coupled to the knitted electrode on the second fabric portion. The other side of the knitted electrode, in contrast, is to be placed in contact with the skin of the user. Then the third fabric portion is folded over to the first fabric portion along the second fold line, so that the connector on the third fabric portion is electrically coupled to the knitted conductive pathway on the first fabric portion and, also, the first fabric portion is located between the second fabric portion and the third fabric portion. Additionally, Fig. 4.19 illustrates the cross-sectional side view of the textile-based electrode system along Line A-A as shown in Fig. 4.17 and Line B-B as shown in Fig. 4.18.

This textile-based electrode system has several advantages over other similar systems. To begin with, the textile-based electrode system can be formed seamlessly in a tubular form (Fig. 4.20), apart from the first two methods of construction mentioned above. Furthermore, the example shown in Fig. 4.20 demonstrates the feasibility of including more than one set of electrodes on the same fabric band; hence, multiple physi-

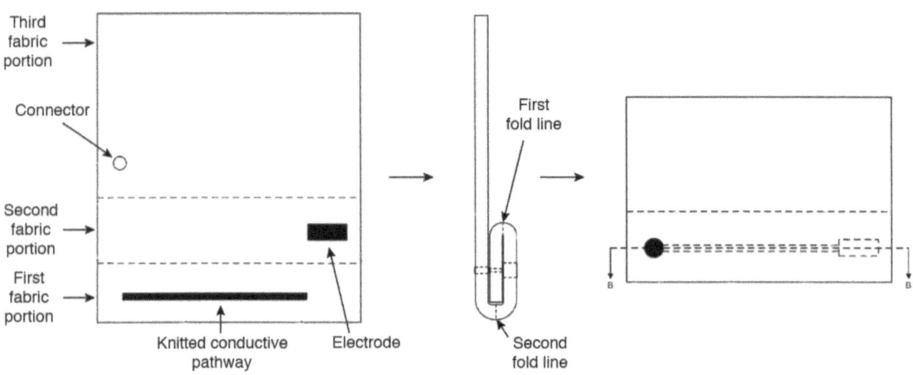

Figure 4.18 Knitted wearable textile-based electrode system: second method of construction with one single piece of fabric.

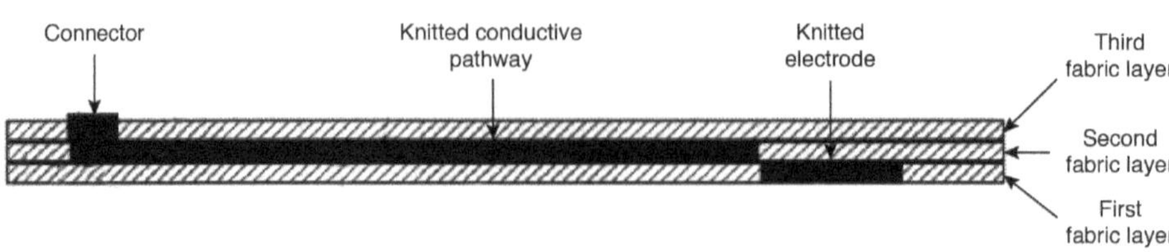

Figure 4.19 Knitted wearable textile-based electrode system: side cross-sectional view along Line A-A as shown in Fig. 4.17 and Line B-B as shown in Fig. 4.18.

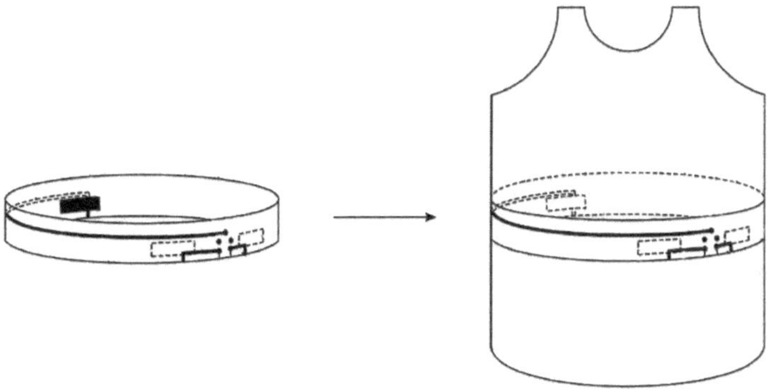

Figure 4.20 Knitted wearable textile-based electrode system: third method of construction with no seams and application to a vest.

ological parameters can be monitored at the same time. In addition, the system can be sewn onto any wearable textile product; for example, a vest as shown in Fig. 4.20, or a waistband, a belt, an arm band, a thigh band, etc. In comparison to the electrodes and conductive pathways that are sewn or stitched onto such systems, the knitted electrodes and conductive pathways have better wear comfort, greater longevity and lower production cost. The reason why this system provides better wear comfort is because knitting can reduce the chafing of the skin which is caused by seams and stitches. A longer life span of the system is ideal as it is more prone to wear and tear upon repeated use and washing. The production cost is lower because only one knitting process is required to produce the entire system, and no additional processes are needed like sewing or stitching the electrodes and conductive pathways onto the fabric. More importantly, the textile-based electrode system is scalable for mass production.

4.5 Conclusion

The esthetics and functional attributes of activewear have been benefiting from rapid developments in fabric technology and garment construction methods. For example, improvements in comfort, handling and esthetic appeal have addressed the concerns of users on the function, protection, weight and appearance of activewear. Early on, high-tech textiles or materials have been applied to activewear because activewear plays an important role in optimizing the performance of athletes. In the future, it is anticipated that the integration of smart technology into activewear will be more prevalent and widely accepted.

References

Berzowska, J., Chanay, F., Menard, S., & Nurkka, M. E. (2013). United States Patent No. US20140343390A1. Available from: https://patents.google.com/patent/US20140343390A1/en?oq=US20140343390A1.

Coltman, C., McGhee, D., & Steele, J. (2015). Bra strap orientations and designs to minimise bra strap discomfort and pressure during sport and exercise in women with large breasts. *Sports Med - Open*, *1*(1), 1–8.

Fildan, G., & Wanzenböck, K. (2004). United States Patent No. US6793556B1. Retrieved from https://patents.google.com/patent/US6793556B1/en?oq=6793556.

Fildan, G., & Wanzenböck, K. (2012). United States Patent No. US8230559B2. Retrieved from https://patents.google.com/patent/US8230559B2/en?oq=US8230559B2.

Fildan, G., and Wanzenböck, K. (2017). United States Patent No. US20170135429A1. Retrieved from https://patents.google.com/patent/US20170135429A1/en?oq=US20170135429A1.

Fildan, G., and Wanzenböck, K. (2018). United States Patent No. US20180092439A1. Available from: https://patents.google.com/patent/US20180092439A1/en?oq=US20180092439A1.

Lin, Y. P., & Zou, Y. C. (2007). China Patent No. CN101205652A.

Liu, M. F. (2009). China Patent No. CN201567416U.

Liu, M. F., & Xie, Y. F. (2010). China Patent No. CN201911358U.

McCann, J. (2005). 4 - Material requirements for the design of performance sportswear. In R. Shishoo (Ed.), *Textiles in sport* (pp. 44–69). Cambridge, England: Woodhead Publishing.

McGhee, D. E., & Steele, J. R. (2010). Optimising breast support in female patients through correct bra fit. A cross-sectional study. *Journal of Science and Medicine in Sport*, *13*(6), 568–572.

McGhee, D. E., Steele, J. R., Zealey, W. J., & Takacs, G. J. (2013). Bra-breast forces generated in women with large breasts while standing and during treadmill running: Implications for sports bra design (Report) (Author abstract). *Applied Ergonomics*, *44*(1), 112.

Shishoo, R. (2005). 1 - Introduction. In R. Shishoo (Ed.), *Textiles in sport* (pp. 1–11). Cambridge, England: Woodhead Publishing.

Voyce, J., Dafniotis, P., & Towlson, S. (2005). 10 - Elastic textiles. In R. Shishoo (Ed.), *Textiles in sport* (pp. 204–230). Cambridge, England: Woodhead Publishing.

Xie, Y. F. (2016). China Patent No. CN106637578B. Available from: https://patents.google.com/patent/CN106637578B/en.

Evaluation of perceived comfort and functional performance of activewear

5

Rong Liu, Elaine Fung, Younus Abida
Institute of Textiles and Clothing, The Hong Kong Polytechnic University, Hung Hom, Hong Kong

5.1 Introduction

Activewear is defined as clothing that is both comfortable and suitable for carrying out sporting activities or exercise (Lexico, 2019). A growing interest in sporting activities and increased awareness of physical health have stimulated and promoted the consumption of activewear and increased expectations and demand. Perceived comfort, as a fundamental and universal feeling, has been considered to be one of the most important attributes in how consumers perceive the acceptability and desirability of functional apparel products in markets worldwide.

Clothing comfort is a complex subjective sensation, which is related to the interactions among materials that come into contact with the body, climatic, and human physiological and psychological factors. Clothing comfort is influenced by a number of qualitative and quantitative variables. Both subjective and objective evaluations that use direct and indirect approaches facilitate the identification of the key variables that influence wear experiences and satisfaction. ASTM, AATCC, and ISO standardized methods are commonly used to directly test the physical-mechanical properties of clothing materials. Nowadays, functional clothing can be used as monitoring tools to measure physiological indicators (e.g., heart rate, blood pressure, body temperature, oxygen consumption, blood lactate concentration, and skeletal muscular activities) to reflect the health and wellbeing of wearers in activities; and clothing also can be used in tracking tests or wear trials to gauge the psychological responses of wearers on-site or after a longer period of use, for example, determining the aesthetic appeal, and snug, heavy, soft, itchy, hot, sticky, or tight/pressure sensations. The materials used for activewear come into direct contact with the skin of the user, thus acting as a near environment of the human body. The receptors in the skin, such as the mechano-, thermo-, and nociceptors, receive a multitude of stimuli through touch, including warmth, cold or pain (Iggo, 1988). A variety of the different interactive properties of materials stimulate these receptors in the skin which result in different perceived comfort levels depending on the intensity of a specific exercise or the climatic conditions (e.g., temperature, relative humidity, radiant heat, and air flow) (Kilinc-Balci, 2011; Choudhury, Majumdar, & Datta, 2011). Liu and Little (2009) established a 5Ps (physical, psychological, physiological, psychophysical and psychophysiological properties) contextual model to evaluate and optimize the comfort and functional

Latest Material and Technological Developments for Activewear. http://dx.doi.org/10.1016/B978-0-12-819492-8.00005-3

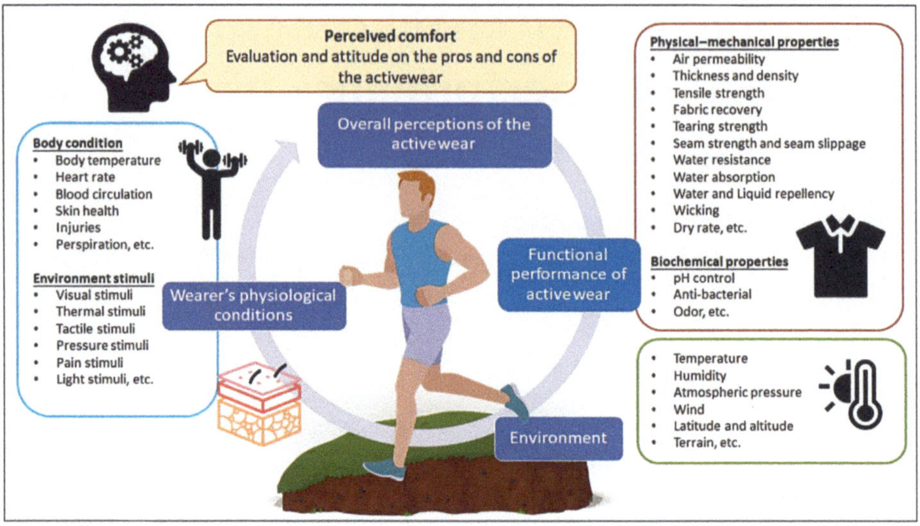

Figure 5.1 Activewear-human body-environment system that reflects interactions between the perceived comfort and material properties (physical, mechanical, and biochemical).

performance of active sportswear. This new design model considers the multiple relationships among wearers, activewear, immediate body space, sports/exercise environment, and wear comfort of users in an active design system, thus contributing to effectively determining developmental strategies that meet the needs of target consumers (Liu & Little, 2009). Fig. 5.1 illustrates an activewear-human body-environment system that involves the aforementioned various clothing properties that may influence the perceived comfort of the wearer (e.g., an athlete). With the development of sensing and electrical technologies as well as computing and informatics, a variety of instruments/devices, methodologies, and algorithms have been explored on their ability to effectively evaluate the functional performances of activewear, which contributes to establishing relationships between perceived comfort and various clothing properties, thus allowing analyses of product designs and their modifications to maximize the wear comfort of consumers.

5.2 Process of perception of comfort

How do humans perceive comfort? The overall perceived comfort is the result of the complex weighting of a combination of inputs from various sensory organs such as the skin, eyes, ears, nose, and mouth toward visual- (color, light, and reflection), physical- (heat and cold), mechanical- (stretching of the muscles, rotation of the joints), chemical- (odorants and acid) stimuli, and so on. The organs send out neurophysiological impulses to the brain through the central neural system for psychological processing, and then a subjective perception is produced toward the stimuli. Following that, a psychological comparative evaluation of the past experience and

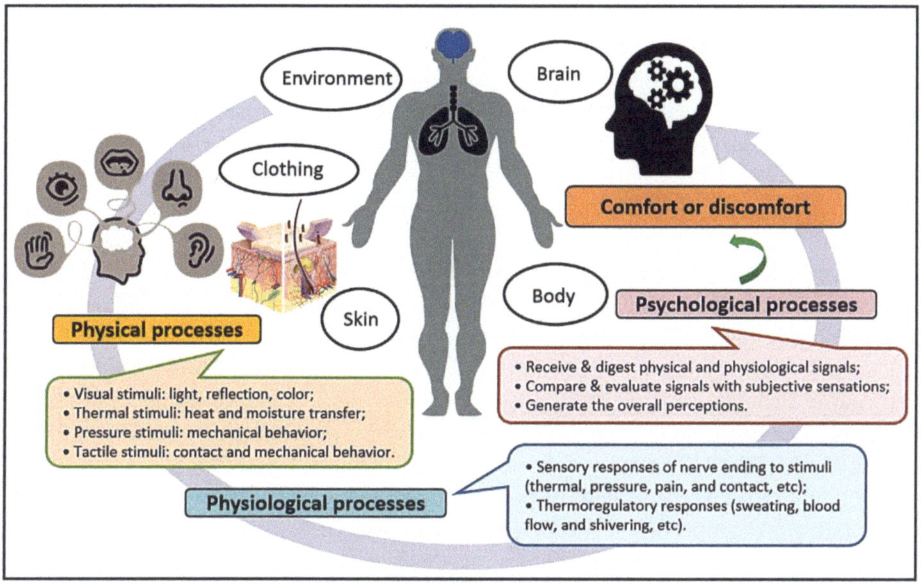

Figure 5.2 Flow of subjective sensory process.

future expectations takes place to form a comprehensive conclusion. The factors that influence subjective judgment could include preferences, expectations, previous experience, and prejudice, etc. (Kamalha, Zeng, Mwasiagi, & Kyatuheire, 2013). Fig. 5.2 illustrates the flow of the subjective sensory process.

The perceived comfort of clothing can also be determined by applying the cognitive load theory (Bell, Cardello, & Schutz, 2005). Cognitive load is the use of the working memory for problem-solving (Kirschner & Paul, 2002). The working memory can be short- or long-term (Baddeley, 1992). According to Bell et al. (2005), the performance of clothing needs to align with the working memory of the wearer. If an activewear user perceives that the wear comfort is low and does not allow her/him to feel comfortable during sporting activities, this will affect her/his short-term working memory and eventually long-term working memory. By examining the cognitive load of activewear performance, activewear users can determine their own set of performance indicators based on the perceived comfort of the activewear.

5.3 Methods of evaluating perceived comfort

Apart from the qualitative data based on the perceived comfort of consumers, quantitative data that are collected through ratings are also crucial for gaining a better understanding on activewear properties and functional performance and improving them. To directly determine subjective response toward activewear, attitude and perception scales can be used to quantitatively measure attitude and perception. The rating scales comprise one or a group of scales that measure a single dimension or multiple dimensions of attitude or perception. The respondent is required to rate an attribute of an object along a continuum with two extremes and descriptors that allow selection of low to high scores or in one of the numerically ordered categories. The rating scales can be designed to focus on a particular (targeted) aspect of an object, or different aspects of an object to reflect the overall attitude (e.g., perceived comfort).

5.3.1 Likert scale

The Likert scale (LS) is widely used in different psychometric studies and psychological measurements to determine attitudes. The LS has been widely used in studies on teaching mathematics to show the relationship between psychometric beliefs and practices (Likert, 1932). As mentioned in Section 5.1, the perceived comfort of activewear is strongly affected by five key properties namely: (1) physical, (2) psychological, (3) physiological, (4) psychophysical, and (5) psychophysiological properties (Liu & Little, 2009). The LS can be used in the form of a question statement paired with a quantitative value to help respondents provide a quantitative value on the five key dimensional properties of perceived comfort (Fig. 5.3). Five ordered response levels are most widely used in the LS (e.g., strongly agree, agree, neutral, disagree, and strongly disagree). After collecting a certain number of responses, a mean score can be calculated for further statistical analysis (Reed, Wolf, Cotton, & Dellon, 2017). For activewear users, more focused performance indicators could be used to assess certain issues of concern, for example, thermo-moisture properties, degree of tightness or amount of pressure, or tactile sensations, etc., thus allowing a better understanding of the factors that contribute to the perceived level of comfort of the users.

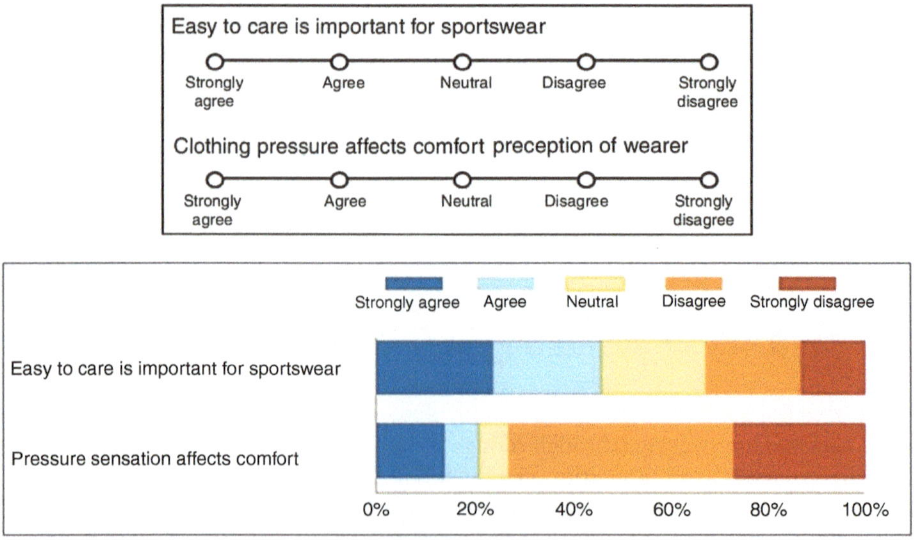

Figure 5.3 Example of Likert scale for activewear assessment.

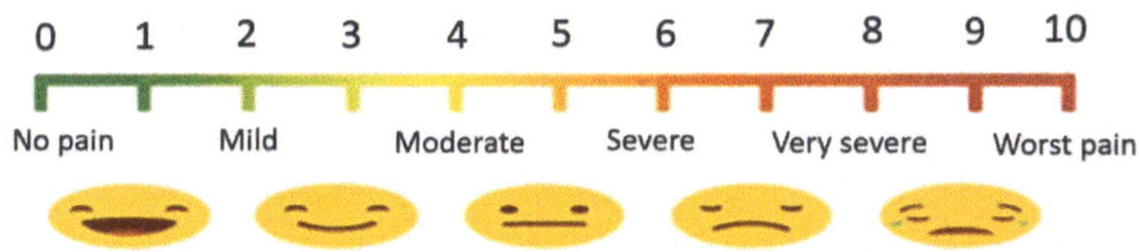

Figure 5.4 Example of visual analogue scale for subjective assessment of pain.

5.3.2 *Visual analogue scale*

The Visual Analogue Scale (VAS) is a unidimensional measure which has been widely applied in studies on adult populations to measure the intensity of attitudes or characteristics that cannot be directly measured. For example, pain is a personal and subjective sensation which is difficult to measure or quantify. The VAS for pain can use a list of descriptors to represent the level of pain. A higher descriptor score of intensity means higher intensity of pain. Four- and five-point VASs are usually applied to assess clinical and treatment outcomes, respectively. Fig. 5.4 is an example of a VAS that consists of a 10 cm line with a statement at each end which represents the extremes of the dimension that is being measured.

5.4 Evaluation of functional performance: fabric level (2D)

The functional performance of activewear can be determined by examining the fabric and clothing levels based on two-dimensional (2D) and three-dimensional (3D) scales. Functional performance could be based on the material properties themselves or the effects of the material properties to determine the exercise performance level and perceived comfort. In this section, the different evaluation methods at the fabric-level are given an introduction, which mainly includes three factors: (1) the physical properties (air permeability, thermal insulation, breathability, thickness, density, weight, and wind resistance), (2) the mechanical properties (elasticity and recovery), and (3) the chemical (biochemical) properties (odor, pH, and resistance to bacteria). Table 5.1 shows the standards used to evaluate both the conventional and high-performance properties of activewear by different organizations globally.

5.4.1 *Physical–mechanical performance*

The physical properties of materials that demonstrate their level of comfort include air permeability, thickness, density, heaviness, heat (thermal) transfer, handle, and flexibility (Kilinc-Balci, 2011). Details on the evaluation methods are discussed below.

5.4.1.1 *Air permeability and thickness*

The air permeability of fabric is one of the major properties that affects the perceived comfort of clothing (Milenković, Škundrić, Sokolović, & Nikolić, 1999), and tested by using an air permeability tester (Fig. 5.5 A) based on ASTM D737-96 Test Method for

Table 5.1 Testing standards of conventional and high-performance properties of activewear

High-performance textile testing	Testing standards					
	ISO	GB, CNS, FZ/T	BS	JIS, JEC	AS, NZS	ASTM, AATCC
Comfortable performance						
Breathability test (water vapor permeability test)		GB/T 12704.1 GB/T 12704.2 CNS 12222 L3223	BS 7209 BS 3424 Part 34 Method 37 BS 3546	JIS L1099 A-1 ($CaCl^2$) JIS L1099 A-2 (water) JIS L1099 B-1/B-2 (Potassium Acetate)		ASTM E96
Deodorization test	ISO 17299-2 (Ammonia) ISO 17299-2 (Acetic acid) ISO 17299-3 (Method A Isovaleric acid)ISO 17299-3 (2-Nonenal)			JEC 301-2015 (Ammonia) JEC 301-2015 (Acetic acid) JEC 301-2015 (Isovaleric acid) JEC 301-2015 (2-Nonenal)		
Moisture management test		GB/T 21655.2 CNS 15659-2 L1038-2				AATCC 195
Quick dry test		GB/T 21655.1, Sec.8.3		JIS L 1096 8.25		AATCC 199 AATCC 200 AATCC 201
Stretch and recovery test		FZ/T 70006 FZ/T 73013 section 5.4.6 FZ/T 73019.2 -section 6.3.6	BS EN 14704-1 Method A&B BS 4294 BS 4952 (replaced by BS EN 14704-1 2005)BS EN 14704-3 Method A	JIS L1018 JIS L1096		ASTM D2594 ASTM D4964 ASTM D6614 ASTM D3107

Protective performance						
Antibacterial test	ISO 22196 ISO 20743	GB/T 20944.2 GB/T 20911.3 FZ/T 73023 -Annex D absorption method FZ/T 73023 -Annex D shaking method		JIS L1902 JIS Z 2801		ASTM E2149 ASTM E2180 AATCC 100
Antistatic test		GB/T 22042	BS EN 1149-1			AATCC 76 AATCC 84
UPF testing - UV protection		GB/T 18830 CNS 15001 L1035	BS EN 13758-1/2 BS 8466 Solar UV properties of hats BS 7914		AS/NZS 4399	ASTM D6544 + D6603 AATCC 183
Water repellency test	ISO 4920	GB/T 4745 CNS 10461 L3202FZ/T 81010	BS EN ISO 24920 BS 3424: Part 26, Method 29D	JIS L1092	AS 2001-2-16-1987	AATCC 22
Water resistance test	ISO 22958 ISO 811 ISO 1420	GB/T 23321 GB/T 4744 CNS 10460 L3201	BS 3424 Part 26, method 29A - static pressure BS EN ISO 20811	JIS L1092 A JIS L1092 B	AS 1066 method 2 (replaced by AS 2001.2.16 -18)	ASTM D751 procedure A (Mullen type tester) AATCC 35: Rain Test AATCC 127
Weathering test	ISO 105 B06 ISO 105 B10 ISO 4892-3	FZ/T 14023 FZ/T 01008 FZ/T 14023 -Annex A				ASTM G155 ASTM G 154 ASTM D4329 ASTM D4587 ASTM D5208 ASTM D4329 AATCC 186
Easy-care performance						
Soil release test		GB/T 30159.1 (only liquid stain) CNS 11309 L3218 FZ/T 01118 FZ/T 24012 Annex A				AATCC 130
Wrinkle free test	ISO 7768 ISO 7769 ISO 9867					AATCC 124 AATCC 128

ISO, International Organization for Standardization; GB, Guobiao Standards; CNS, FT/Z, Chinese National Standards; BS, British Standards ; JIS, Japanese Industrial Standards; JEC, Japanese Electrotechnical Committee; ASTM, American Society for Testing and Materials Standards; AATCC, American Association of Textile Chemists and Colorists.

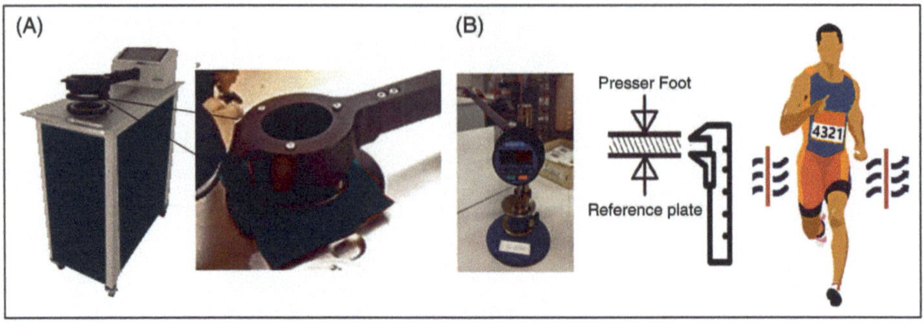

Figure 5.5 (A) Air permeability tester and (B) standard thickness tester.

Air Permeability of Textile Fabrics, or BS EN ISO 9237 Textiles-Determination of the permeability of fabrics to air, etc. According to the American Association of Textile Chemists and Colorists (AATCC): "the sample is secured on the permeability tester. The tester then generates an air pressure gradient across both sides of the sample". The air permeability of the sample can be determined by measuring the rate of the degeneration of the pressure gradient. A lower rate of degeneration of the pressure gradient means lower air permeability.

Fabric thickness greatly affects air permeability. By measuring and recording the distance between a parallel presser foot which provides the pressure and a reference plate on which the fabric is placed on a standard thickness tester, the mean of the fabric thickness can be calculated and displayed (Fig. 5.5B).

5.4.1.2 Tensile strength

Fit is becoming a primary criterion for form-fitting sportswear apparel (Ashdown, 2017). Sportswear fit can be however affected by the body movement, body proportions, size, and elasticity of the fabrics (LaBat & DeLong, 1990). Tensile strength is one of the methods to evaluate the elasticity of the activewear fabrics. The tensile strength of a material is the maximum amount of tensile stress that it can take before failure, for example breaking. The tensile strength of elastic fabric can be measured by using a CRE universal strength tester (Fig. 5.6A). By holding one end of the fabric in place and extending the other end with a movable jaw, the force of the extension (at a constant rate) and the point of breakage during elongation can be measured and recorded.

5.4.1.3 Fabric recovery

During sporting activities, the muscles are being stretched at a high movement frequency. The movement of the body therefore expands the skin by 10%–50% (Senthilkumar & Anbumani, 2011). To improve sports performance, elastic fabrics have been widely used in activewear because they not only facilitate body stretching in movement, but also can resist stretching through providing a high rate of recovery, which allows the fabric to return to its original size and shape after repetitive physical movements are carried out (Shishoo, 2005). The recovery of elastic fabric can be

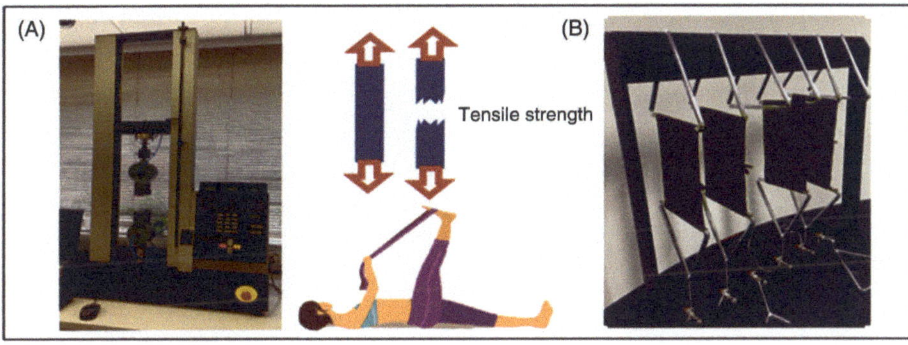

Figure 5.6 (A) CRE Universal strength tester and (B) knitted fabric stretch recovery tester.

classified into two types (Senthilkumar & Anbumani, 2011): (1) dynamic elastic recovery (DER) and (2) static elastic recovery (SER). DER refers to how sportswear instantaneously responds to body movement during sporting activities while SER refers to the dimensional stability of sportswear. By analyzing the DER and SER, the appropriate fabric can be used as activewear to enhance sports performance. One of the most commonly used methods to measure the recovery of activewear fabric is ASTM D 2594-99A Standard Test Method for Stretch Properties of Knitted Fabrics Having Low Power. Using a knitted fabric stretch recovery tester (Fig. 5.6B), elastic knit fabric can be stretched at a specific tension. The extension and recovery of the fabric can be measured and recorded at the same time (Millet, Perrey, Divert, & Foissac, 2006).

5.4.1.4 Tearing strength

Because of the intense movement during sporting activities, the activewear fabric is extended and elongated at the same time, which may cause the fabric to tear. By using an Elmendorf tester (Fig. 5.7A) and a universal strength tester (Fig. 5.7B), the maximum force which is required to continuously tear a piece of pre-cut fabric can be

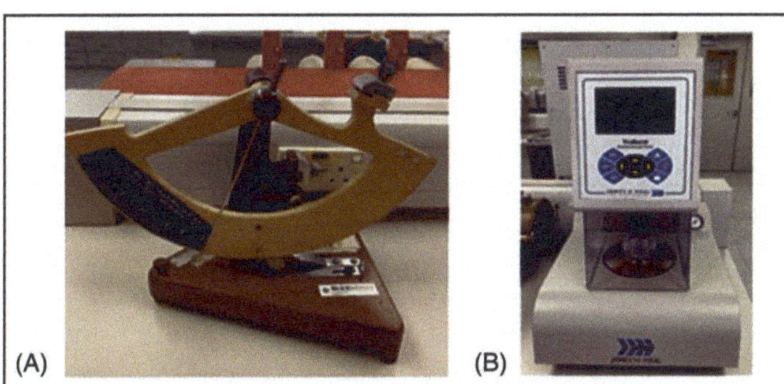

Figure 5.7 (A) Elmendorf tester and (B) universal strength tester.

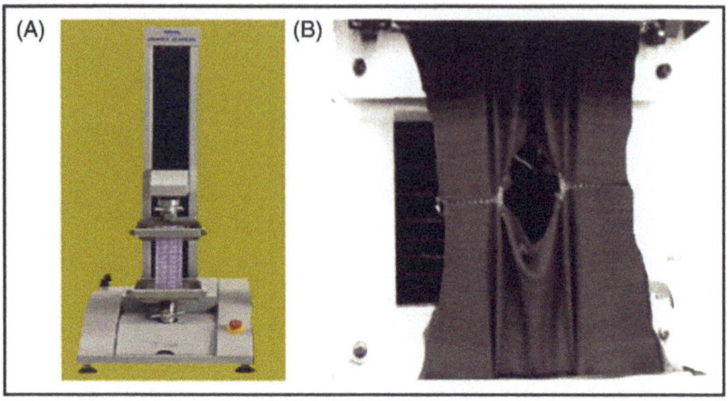

Figure 5.8 (A) Seam slippage tester and (B) the tested fabric.

measured and recorded. The recorded force implies the tearing strength of the tested activewear fabric.

5.4.1.5 Seam strength and seam slippage

Seam quality is one of the most important factors that affects the durability of garments (Shi, Zhang, & Wang, 2017), especially activewear which endures intense physical movement during exercise. The maximum force which needs to break the seam is recorded as the seam strength, which can be tested by using a CRE universal strength tester as shown in Fig. 5.6A. Fig. 5.8 A shows a seam slippage tester. The seam of the activewear is placed midway between two gripping jaws and then stretched. The seam slippage is measured by the force that displaces the seam without breaking the seam.

5.4.2 Thermo-moisture properties and evaluation

The moisture transmission of activewear apparel greatly affects the sensorial and thermo-physiological comfort of the wearer. According to Hes and Williams (2011), if clothing can balance the thermoregulation of the body and maintain a constant body temperature of 37°C, it is considered to be a garment that allows the wearer to be comfortable. This is because a constant body temperature means that a constant amount of heat is applied to the body while extra heat will transfer from the body to the outside of the garment (Farnworth & Dolhan, 1984). Body temperature is greatly affected by the moisture management properties of activewear, which directly relates to the level of comfort perceived during intensive sporting activities. After intense activity, moisture is transferred from the surface of the skin to the outside environment. Then perspiration can easily evaporate and enhance thermoregulation of the body (Suganthi & Senthilkumar, 2018). A moisture management index (MMI) is commonly used to determine the moisture management properties of fabrics. Determining the MMI is based on five test parameters, namely: (1) water resistance, (2) water absorption, (3) water repellency, (4) fiber and yarn wicking properties, and (5) drying rate.

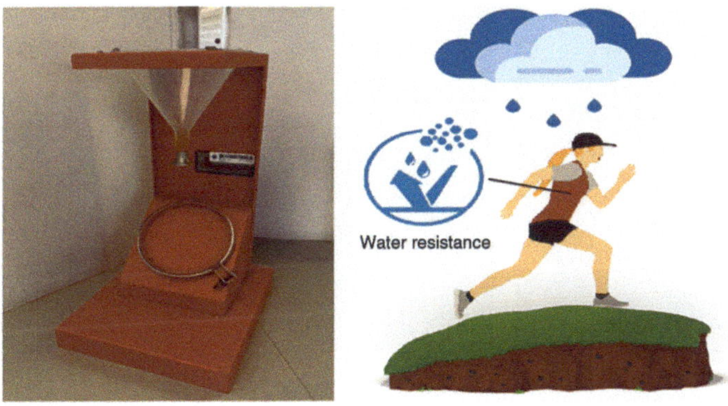

Figure 5.9 Rain tester.

5.4.2.1 *Water resistance*

The water resistance of activewear fabric can be tested by using AATCC 35 Water Resistance: Rain Test. By using a rain tester (Fig. 5.9), water is applied onto the surface of activewear fabric, and blotting paper is mounted behind the tested fabric during the test for a particular period of time. Then the volume of water that penetrates through the fabric and absorbed by the blotting paper can be measured. The volume of water that is absorbed by the blotting paper indicates the water resistance level of the fabric. Additionally, the water resistance of activewear fabric can be measured by using AATCC 22 Water Repellency-Spray Test. By using gravity force, the spray test applies water to the surface of the activewear fabric for a specific amount of time. Then the tested fabric can be assessed based on the wetted surface area.

5.4.2.2 *Water absorption*

The propensity of activewear fabric to absorb water can be measured by using AATCC TM 79 Absorbency of Textiles or ASTM Test Method TS-018 Procedure for Absorbency. The test is conducted by placing the sample fabric in an embroidery hoop to remove all of the creases. A drop of water is dripped onto the fabric surface by using a burette at a distance of 9.5 mm. Then the time required for the fabric to completely absorb the water droplet is recorded.

5.4.2.3 *Water and liquid repellency*

Water repellency can be tested by using AATCC 127 Water Resistance: Hydrostatic Pressure Test (Fig. 5.10). This test is conducted by dripping 5 drops of water onto the surface of the tested fabric and measuring the pressure that is required to force the water droplet to pass through the fabric. The amount of liquid mixture absorbed by the fabric is measured to determine the repellency by using AATCC 193 Aqueous Liquid Repellency: Water/Alcohol Solution Resistance Test. The liquid used in the test is a combination of alcohol and water. The time that it takes for the fabric to absorb the

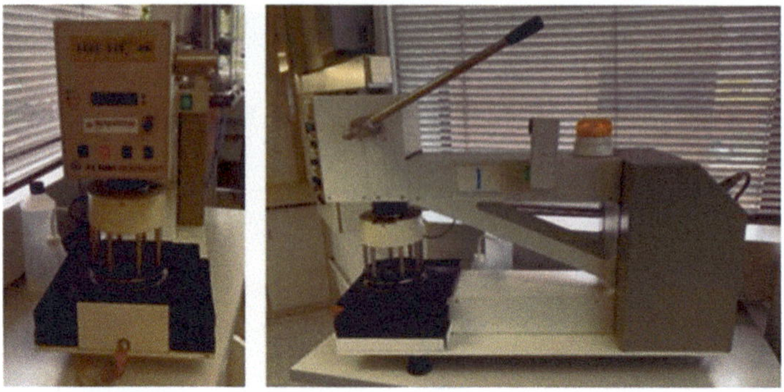

Figure 5.10 Hydrostatic head tester.

liquid mixture is recorded, and meanwhile, the liquid which has not been absorbed into the fabric is given a rating.

5.4.2.4 Wicking

Activewear fabric needs to have the ability to move liquid moisture from the surface of the skin to the surface of the fabric after intensive physical activity to enhance the comfort of the wearer (Fig. 5.11). This is called wicking, in which moisture can move from the capillaries of the fabric to its surface. By measuring the rate of wicking, the ability of a fabric to remove perspiration from the surface of the skin to the surface of the fabric can be determined. Different tests are used to measure the rate of wicking; namely: (1) AATCC TM 197 Vertical Wicking of Textiles (gravity involved), and (2) AATCC TM 198 Horizontal Wicking of Textiles (no gravity involved). In the time required for the liquid to move a given distance on the fabric and the distance that the liquid has moved on the fabric at a given time is measured by using these two tests. By evaluating the collected data, the ability of the fabric to transport moisture from one point to another can be determined. The shorter the time to transport the liquid moisture for a certain distance, which means that the tested activewear fabric has a higher wicking ability.

Figure 5.11 Wicking of sweat during physical activity.

5.4.2.5 *Drying rate*

The drying rate refers to the drying ability of the activewear fabric. Even if the fabric has a high wicking ability, the moisture will be retained in the garment which causes the garment to feel damp and affects the perceived comfort if the garment cannot easily dry by itself. To measure the drying properties of activewear fabric, the standard methods of AATCC TM 199 Drying Time of Textiles: Moisture Analyzer Method and AATCC TM 200, 201 Drying Rate of Textiles: Heated Plate Method can be applied. By using a gravimetric moisture analyzer, the drying time of fabric at an evaluated temperature, which includes the temperature of the surface of the skin (37°C), is measured. The drying rate of the textile test measures the change in fabric weight after the fabric is wetted with water. AATCC TM 201 uses a heated plate testing method by which the activewear fabric is saturated in a specific amount of water (2.5–3 times that of the fabric weight) and placed onto a hot plate which is the same as the body temperature (37°C). The fabric is then weighed on a scale with a mesh holder every 5 min. Once the fabric weight is reduced and less than 105% of its original weight, the fabric is considered to be dry, and the weight recorded over time is the drying rate.

5.4.2.6 *Thermal resistance*

Thermal properties of textiles closely relate to the thermal perception of the user (Tessier, 2018; Beaudette & Park, 2017). Thermal resistance, as one of the thermal properties, can be determined by measuring the difference in the temperature between both sides of the material in connection with human skin (ISO 11092; 2014). By using a sweating guarded hotplate, the heat transfer process between the human skin and the material can be simulated (Testex textile 2020). The sweating guarded hotplate includes a metal plate that is surrounded by a thermal guard. As the temperature of the bottom of the protective plate remains constant, the heat loss can only be generated through the specimen. Once the testing condition reaches the stable level, the temperature of the air layer above the specimen surface and the surface of the specimen will be measured (Huang 2006). Then the thermal resistance can be determined by using the thermal resistance formula, which refers to the ratio between the heat flux per unit area to the specimen to the temperature difference on the specimens' both sides (ISO 11092; 2014; Testex textile 2020).

5.4.3 Chemical (or biochemical) properties and evaluation

Most functional textiles are treated with a chemical (biochemical) finishing to improve their wear performance. Chemical (biochemical) tests can be conducted to help meet the standard configurations and address the potential problems during the production stage and with customer use.

5.4.3.1 Fibre composition

Fibers used in textile production can be classified into synthetic and natural fibers. The latter can be further classified into protein fibers (animal fibers) and cellulose fibers (plant-based fibers). To identify the fibers used for activewear, AATCC TM20 Fiber Analysis: Qualitative and TM20A Fiber Analysis: Quantitative can be applied. The former is a visual examination, that is, using a microscope to visually identify the fibers based on their longitudinal and cross sections. The latter is a chemical test that involves burning and staining the fibers and determining the solubility. Table 5.2 shows the solubility of synthetic fibers in solvents under different conditions (Fan, 2008).

5.4.3.2 pH value

Acidic chemicals may remain on activewear fabric after chemical treatment, which affects the pH value of the fabric and cause skin allergies when the fabric comes into contact with the body. AATCC TM81 pH of the Water-Extract from Bleached Textiles can be used (AATCC, 2019) to test if the acidic/alkali content has been properly eliminated from the activewear fabric. Here, a 10 ± 0.1 g sample of an activewear fabric is tested as an example. 250 mL of distilled water is boiled for 10 min, and the sample is placed into the boiling water and boiled for 10 more minutes. Then the sample and water are allowed to cool down to room temperature in a covered beaker. Finally, a tweezer is used to remove the sample, with any excess liquid dripped back into the beaker. Then the pH value of the liquid in the beaker can be tested.

5.4.3.3 Anti-bacteria properties

Resistance to bacteria is one of the most important factors for the comfort performance of apparel. Textile used for clothing is a good medium for the growth of bacteria, especially tightly fitting activewear (Deng, Wang, & Wang, 2016). Perspiration will generate an unpleasant odor and affects the perceived comfort of activewear. Bacteria can also lead to different skin diseases which affect the health of the wearer (Wang, 2018; Zhao & Wang, 2013). To test the ability of a fabric to resist bacteria, the sample needs to be inoculated with test organisms. The bacteria will be eluted from the sample after incubation by shaking the sample in a neutralized solution. Then, to determine the anti-bacteria properties, the bacteria count is determined. A greater reduction in bacteria means better resistance of the fabric to bacteria.

Table 5.2 The solubility of synthetic fibers in solvents

Active-wear synthetic fibers	5 min, 20°C, 85% formic acid	5 min, 139°C, m-cresol (glycerol bath)	10 min, 90°C, Dimethylformamide
Nylon 6	Soluble	Soluble	Soluble
Nylon 66	Soluble	Soluble	Insoluble
Acrylic	Insoluble	Mass formed plastic	Insoluble
Polyester	Insoluble	Soluble	Soluble

Figure 5.12 Odor testing.

5.4.3.4 Odor testing

Odor is the scent or smell of humans that is caused by volatilized chemical compounds. As sporting is a highly intensive activity, body heat will be generated during movement. To regulate the body temperature, the human body produces sweat. Sweat will move from the inner skin layer to the outer skin surface through the sweat glands. Then sweat will produce volatilized chemical compounds and an unpleasant odor. To deal with this problem, different studies have examined the systems and methods for odor absorption (Sesselmann, 1996, 2000). Here, we carry out a test to determine the odor properties of fabric (Fig. 5.12). A sample fabric (40 g) was placed above a sodium carbonate solution (300 mL) in a closed container. Then, the closed container is placed into an oven at a temperature of 37 °C ± 2 °C for 15 h. A minimum of six individuals are asked to independently evaluate and rate the intensity of the odor in accordance with five intensity grades; namely Grade 1 – odorless; Grade 2 – weak odor; Grade 3 – tolerable odor; Grade 4 – annoying odor, and Grade 5 – intolerable odor. The more detailed testing method and procedure can refer to SNV 195651: Determination of the development of smells of finishings (sensory test), etc.

5.5 Evaluation of functional performances: clothing level (3D)

The materials and functions of activewear have impacts on the performance of dynamic wear. Laboratory testing of instrumental materials allow testing to be carried out in a controlled environment and increases the ease of reproducing the testing results more quickly at a lower cost. However, the results derived from a laboratory apparatus at the fabric level may not truly reflect the overall comfort, physiological benefits, and biomechanical behaviors of activewear. Therefore, different manikins and wear trials have been used to examine the effects of 3D dynamic wear. Thermal-, sweating- and pressure- manikins are the most common types of manikins used in objective assessments of garments. Trials with human subjects are usually conducted in a controllable chamber or laboratory room in ambient conditions and controlled temperature, humidity, air flow rate, etc. Treadmills or bicycle ergometers are frequently used to test activewear. Probes or sensors that are embedded in the manikins or attached to the body of the tested subject allow a large amount of physical, mechanical, and physiological data to be collected, such as thermal insulation, moisture accumulation, water vapor resistance, breathability, heat loss, pressure magnitude, skin temperature, skin moisture, heart rate (HR), blood pressure (BP), epidermal water content, oxygen consumption, and blood flow, etc.

5.5.1 Manikin assessment in climatic chamber

The use of static or moveable sweating/thermal manikins in fundamental research and development of high-performance clothing has grown rapidly over the years. Heat dissipation, evaporative heat flow, constant skin temperature, heat strain risks, etc., can be simulated by using sweating/thermal manikins to estimate the effects of different conditions (environmental temperature, exercise level, clothing structure, and material properties) on the physiological response and perceived comfort of the human body (DenHartog & Deaton, 2015). Thermal manikins are commonly made of copper, plastic or waterproof fabrics, with a controllable heating/sweating subsystem, and data measurement and analyzing subsystems. Some of the most well-recognized sweating/thermal manikins include Coppelius (a Finnish manikin), TARO and KEM (Japanese manikins), Walter (Hong Kong manikin), SAM (Swiss manikin), and Newton and ADAM (American manikins) (Holmér, 2004; Kilinc-Balci, 2011). Table 5.3 shows the international testing standards that have been developed for evaluating thermal comfort by using a manikin.

In addition to sweating/thermal manikins, static or dynamic pressure manikins have been also developed to directly test the pressure behavior of elastic or tight-fitting clothing. Multiple pressure sensors are incorporated onto different sites of the manikin surface to sense the pressure imposed by a garment and the collected data are then simultaneously transferred to an acquisition device. To more realistically simulate the body shape, the manikin can also extend transversely in the anterior-medial and posterior-medial directions to a specific distance (e.g. 5 cm) (Wang, Cui, Zhang, & Feng, 2011; Yu, Fan, Qiao, & Tao, 2004). These manikin systems are a convenient resource for evaluating pressure, sweating, and the thermal properties of activewear, thus laying the foundations for functional clothing design and product development in industry practices.

5.5.2 Psycho-physiological evaluation in human trials

Wear trials are time- and work- intensive. To obtain reliable statistics, a number of tests must be carried out, and a large amount of data needs to be analyzed with the appropriate statistical methods. The challenges of wear trials are factors related to

Table 5.3 Thermal manikin related testing standards.

Standards	Description
ASTM F1291	Standard Test Method for Measuring the Thermal Insulation of Clothing Using a Heated Manikin
ASTM F2370	Standard Test Method for Measuring the Evaporative Resistance of Clothing Using a Sweating Manikin
ASTM F1720	Standard Test Method for Measuring Thermal Insulation of Sleeping Bags Using a Heated Manikin
ASHRAE 3739	Evaluating Thermal Environments by Using a Thermal Manikin with Controlled Skin Surface Temperature
EN-ISO 15831	Thermal Manikin for Measuring the Resultant Basic Thermal Insulation
ISO 7920	Estimation of The Thermal Characteristics of Clothing
ENV 342	Protective Clothing against Cold
ASHRAE HI-02-17-4	Measurement of Thermal Comfort and Local Discomfort by a Thermal Manikin

the subjects, including their attendance, health, weight/fitting stability, emotions, etc. These factors may influence to some extent the reproducibility and consistency of the results. Here, a pilot study illustrates and examines the effects of different types of activewear on the physiological characteristics of young athletes (Liu, Little, & William, 2012, 2014).

A study case

Five physically fit male participants (with a mean age of 22.4 years old, height of 179.32 cm, weight of 64.05 kg and BMI of 19.95 kg/m^2) took part in the study. The functional performance of four sets of natural and synthetic active running and cross-training suits was assessed. These four sets of suits were selected from different athletic suits based on the testing results of the "Walter" sweating manikin. Coolmax fibers, Meryl microfibernylon, polyester treated antimicrobial agent for odor control, and thermoplastic polyurethane (TPU) supporting bands were used in the active suits (Table 5.4). A series of different physiological parameters (HR, BP, blood

Table 5.4 Basic characteristics of the tested activewear

Sample	Material and design features	Fabric properties
CFA-1	Coolmax polyester/Lycra, full-body garment with support panel	D: 3886 stitches/in^2) on top, 3762 on bottom; W: 291.02 g.
CFA-2	Meryl microfiber/Lycra, full-body garment with gradient pressure design	D: 3886 stitches/in^2) on top, 3968 on bottom; W: 336.55 g.
CFA-3	Polyester/elastane, full-body garment, lightweight and good moisture management	D: 3355 stitches/in^2) on top, 3036 on bottom; W: 282.33 g.
CFA-4	Nylon/spandex, full-body garment with TPU supportive web	D: 2450 stitches/in^2) on top, 3710 on bottom; W: 392.86 g.

D, density in main body, stitches/in^2; W, weight in gram, g.

oxygen level SpO_2, myocardial oxygen consumption DP, skin temperature T_{sk}, and garment temperature $G_{gn,}$) were evaluated during the different stages of exercise (beginning, middle, rest, and end) and based on different exercise intensities (3.5 miles/h, 6 miles/h, and 8 miles/h). The skin pressure as a biomechanical indicator was evaluated on 16 areas of the participant by using multi-channel FlexiForce pressure sensors and a pressure monitoring system. The experimental protocol is shown in Table 5.5.

Fig. 5.13 A and B illustrated the part of experimental settings (treadmill and SpO_2 probes) and the technical face and back views of the studied four active suits in the wear trials. The evaluation results indicated that the use of form-fitting active suits maintains the body in a normal SpO_2 range, and produces a lower DP, HR and T_{sk} compared to those in the control condition. Fig. 5.14 A presented the variation profiles of the tested HR data of the four tested suits and control condition. The produced physiological and psychological responses are related to the physical-mechanical and transmission properties, and fitting dimensions of the active suits. The changes in exercise intensity also affected the physiology of the male participant and their immediate near environment. The studied active suits delivered an interface pressure that ranges from 1679.88 to 2752.89 Pa onto the skin (Fig. 5.14 B), and sample suits CFA-2 and CFA-3 with a gradient pressure design, light in weight and moisture-thermoregulation function showed a relatively consistent performance at pre- and post- experiment, and rated a greater degree of perceived comfort by the participants after the experiment.

Table 5.5 Physiological experimental protocols

Period (min)	Accumulated time (min)	Activity	Physiological measurements
10	0–10	Change into the studied suits	Pre-experiment body weight, garment weight
20	0–20	Rest, stretching	HR, BP, SpO_2, DP, T_{sk}, T_{ga}, CS
5	20–25	Walking (3.5 miles/hr)	HR, SpO_2,
1	25–26	Rest 1 (after walking)	HR, BP, SpO_2, DP, T_{sk}, T_{ga}, CS
5	26–31	Running (6 miles/hr)	HR, SpO_2
[a]	31–32	Mid-run test	HR, BP, SpO_2, DP
5	31–36	Running (6 miles/hr)	HR, SpO_2
1	37–38	Rest 2 (after running)	HR, BP, SpO_2, DP, T_{sk}, T_{ga}, CS
5	38–43	Running (8 miles/hr)	HR, SpO_2
[a]	43–44	Mid-run test	HR, BP, SpO_2, DP
5	44–48	Running (8 miles/hr)	HR, SpO_2
1	48–49	Rest 3 (after running)	HR, BP, SpO_2, DP, T_{sk}, T_{ga}, CS
5	49–54	Walking (3.5 miles/hr)	HR, SpO_2
10	54–64	Rest, stretching	HR, BP, SpO_2, DP, T_{sk}, T_{ga}, CS
10	64–74	Remove suits	Post-experiment body weight, garment weight

[a] Measurements while running. HR, heart rate; SpO_2, Blood oxygen level; BP, blood pressure; DP double product (myocardial consumption); T_{sk}, skin temperature; T_{ga}, garment temperature; CS, comfort survey.

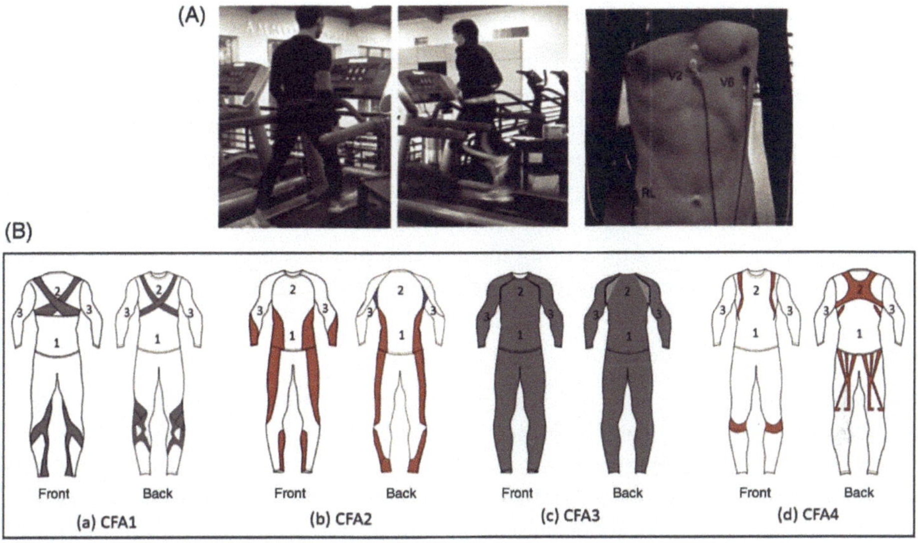

Figure 5.13 (A) The settings of the wear trials and (B) the studied active suits (Liu, Little, William, 2012; 2014).

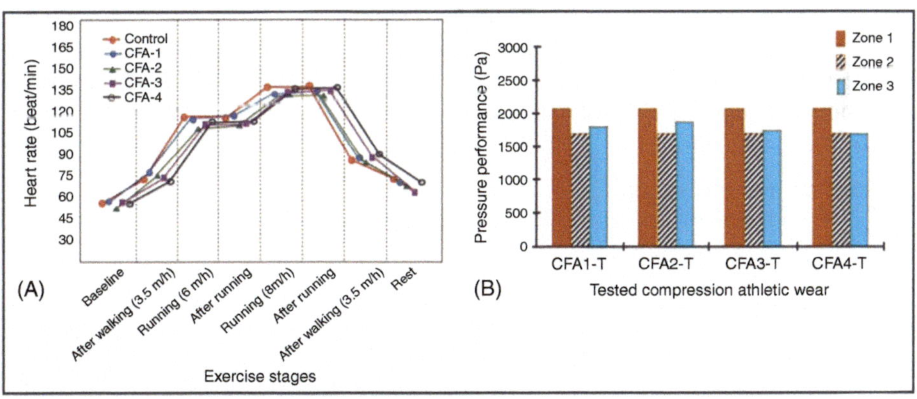

Figure 5.14 (A) Heart rate test and (B) skin pressure test in active suits wear trials.

In addition, more physiological parameters were assessed to study the functional performance of activewear. For example, Lee, Hyun, and Tokura (2000) evaluated the effects of tight clothing on the circadian rhythms of core temperature and salivary melatonin. Liu et al. (2008, 2012, 2019) evaluated the effects of compression textiles and garment on blood flow and vascular function of the lower limbs and resting salivary cortisol and urinary catecholamine excretion in women. Miyasuji et al. (2002) studied the impact of pressure of different types of brassieres on the autonomic nervous system activity by using a power spectral analysis of the heart rate variability.

5.5.3 Biomechanical evaluation in wear trials

With more awareness around the importance of health, biomechanical elements have been increasingly incorporated into activewear and accessories to improve support to the joints, muscle activity, and body movement. These factors not only influence the perceived comfort, but also ergonomic comfort, which determines fit and freedom of movement. Fabric morphology, fabric elasticity, stiffness, and pattern construction (Hawley, 2000) are closely related to the biomechanical performances of activewear, which can be evaluated by measuring the clothing pressure or level of body rotation, and by applying video-based body motion capture analysis, etc.

5.5.3.1 Pressure evaluation

The development of the most efficient and reliable pressure sensors that are high in accuracy is a substantial challenge in the field of function assessment. Variations which include a complex contact surface, stretchable fabric and tissue deformation mean that pressure sensors need to have a high degree of flexibility and reliability to obtain consistent results in repeated tests (Tyler, 2018). There are different kinds of pressure sensors for assessing interface pressure. For example, Liu et al. (2005, 2007, 2014) conducted a series of pressure assessments on compression stockings and active sportswear by using the Flexiforce tactile pressure sensors, air-filled pressure sensors, and self-developed multi-channel signal systems. Flexiforce pressure sensors (Tekscan) are piezoelectric, ultra-thin (0.1 mm or 0.004 in), and flexible (Fig. 5.15A). Its insertion produces less influence at contact condition. Lai & Li-Tsang, (2009) applied Pliance X system (Fig. 5.15B) to measure pressure at pressure garment-scar interface and conducted wear trials to validate its applicability. More types of pressure sensing devices are listed below:

(i) Pneumatic pressure sensors
- Oxford Pressure Monitor – Talley Ltd., Ramsey, Hampshire, UK.

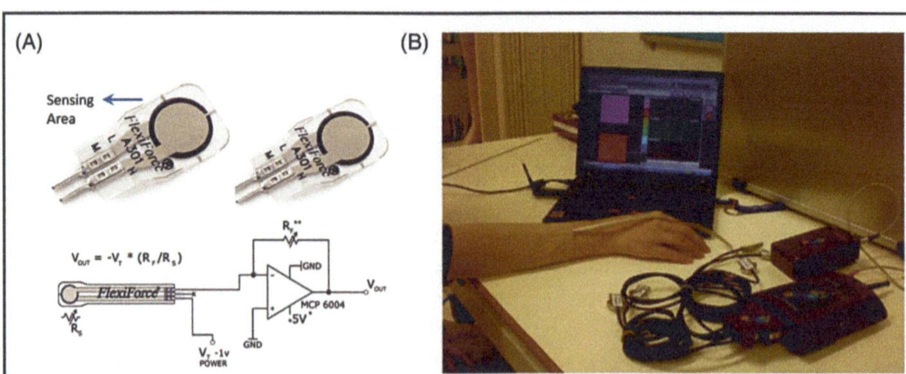

Figure 5.15 Pressure sensors: (A) piezoelectric pressure sensor and (B) Pliance pressure sensor.

- Digital Interface Pressure Evaluator – Next Generation Corporation, Temecula, CA., USA.
- MST MKIII Salzmann – Salzmann Medico, St Gallen, Switzerland.
- Kikuhime – Meditrade, Soro, Denmark.

(ii) Capacitive pressure sensors

- Kulite XTM190 – Kulite Semiconductor Products, Leonia, NJ., USA.
- Precision – Precision Measurement Corporation, Ann Arbor, MI., USA.
- Xsensor – Crown Therapeutics, Belleville, IL., USA.
- Pliance – Novel Munich, Germany.

(iii) Piezoelectric or fluid-filled pressure sensors

- MCDM-1 – Mammendorfer Inst. Physik, Munich, Germany.
- Flexiforce – Tekscan, South Boston, MA., USA.
- Strathclyde Pressure Monitor – University of Strathclyde, Scotland.
- Skip Air Pack Analyzer – AMI Corporation, Japan.

(iv) Resistive and strain gauge pressure sensors

- Fscan, Iscan – Tekscan, South Boston, MA., USA.
- Rincoe Socket Fitting System (SFS) – Rincoe and Associates, Golden, CO., USA.
- MCDM - Mammendorfer Inst. Physik, Munich, Germany.
- Dia-Stron – Dia-Stron Ltd. Andover, Hampshire, UK.

Each pressure sensor listed here has advantages and also some limitations as summarized in Table 5.6.

5.5.3.2 Motion capture analysis

Motion capture analysis (MCA) systems have been used to analyze motion during walking, running, sprinting, landing, cycling, etc. (Fig. 5.16). Motion capture of exercise/sporting performance aims to track and record the movement of athletes in real time to determine their physical conditions, athletic performance (Marqués, Cela, & Gisbert, 2017), technical expertise, and injury process, as well as for prevention and rehabilitation purposes. The use of activewear may facilitate, enhance, or restrict the movement of athletes during dynamic wear. MCA can therefore be used to measure the position of the wearer and movement orientation in a physical space and record the information, thus indirectly estimating the functional performance of activewear. In the recent two decades, motion capture technologies have become a cutting-edge

Table 5.6 Characteristics of different pressure sensors

Types	Advantages	Disadvantages
Pneumatic	Thin and flexible probes Relatively inexpensive and easy to use	Dynamic measurements need extra equipment Temperature sensitive
Fluid filled	Flexible and able to take dynamic measurements	Thick when filled with fluid and causes problems during movement
Resistive	Thin sensors that can take dynamic measurements	Sensitive to 3D contours and stiff

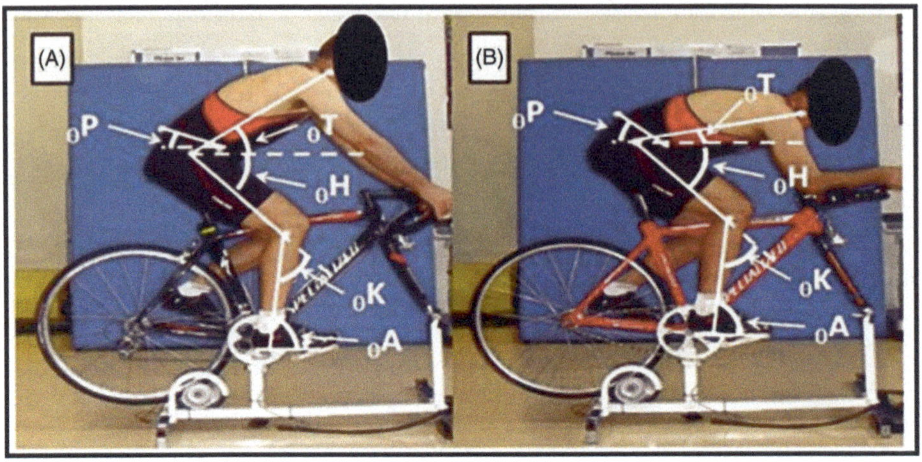

Figure 5.16 Motion capture analysis of cycling exercise: (A) road cycling posture and (B) triathlon/time trial posture (Bini and Carpes, 2017).

research area in which motion data are analyzed and products are developed for a variety of practical applications. Currently, the more elite activewear designs use MCA to optimize the garment structure and function, thus enhancing the sports performance of the wearer during intense competitive events.

Motion capture systems commonly involve several steps as summarized below (Naugle, 1999):

- Set up cameras in a studio, in which the cameras are placed in different locations for multiple captures;
- the captured locations and camera locations are calibrated;
- the cameras capture the motion and data are collected; and
- the data are analyzed to construct 2D and 3D objects.

Table 5.7 outlines the existing motion capture systems and their specifications for body movement studies. Specifically, the optoelectronic measurement system (OMS) is a highly accurate MCA. Optitrack and VICON are commonly used equipment in an OMS (Corazza, Mündermann, Gambaretto, & Ferrigno, 2010). In OMSs, several cameras are placed in specific positions around the test object (here, it is the wearer of activewear) to detect movement from different angles. Each camera is equipped with a strobe (e.g., infra-red strobe) to illuminate the object of testing. Reflective markers are adhered onto the subject who is being tested at different anatomical points. The light that emerges from the camera is reflected by the reflective markers back to the sensors of the cameras. The sensors receive the light signals and create a contrasting image to estimate the 3D position of the reflective marker by using the time-of-flight and triangulation methods. The time-of-flight is the time taken by the light to travel the distance from the origin (camera strobe) to an object (reflective marker) through a medium. The triangulation method uses laser beams that are produced by three sources that are spatially distributed. The accuracy of OMSs depends on the

Table 5.7 Motion capture systems and specifications

System	Marker	Environ-ment	Marker weight	Marker nature	No. of Cameras	Capture volume/ camera	Sample frequency
OMS Opto-track	Active	Indoors	<10 g	IR LED	8	3.6 × 2.6 × 3.7 m	3500 Hz
VICON 460	Passive	Indoors	<10 g	Reflective	6	Depends on lens	2000 Hz
VICON T-40	Passive	Underwater	<10 g	Reflective	10	Depends on lens	2000 Hz
VICON MX-13 MX-40	Passive	Indoors	<10 g	Reflective	>24	Depends on lens	2000 Hz
iGPS	Active	Indoors	>30 g		Infinity	55 m (circular)	50 Hz
EMS WASP	Active	Indoors	-	-	-	-	125Hz

experimental set-up. The primary factors that may affect the accuracy of OMSs are (Maletsky et al., 2007):

- Resolution of each individual camera;
- Positions of cameras relative to each other;
- Distance between each camera;
- Distance between camera and reflective marker;
- Type, number and position of all of the reflective markers; and
- Motion capture volume - the marked area of each reflective marker within which motion capture data are received.

The capture volume depends on the number of cameras and the field of view of each camera. The VICON MX-13 measurement system has been reported to have the largest measurement range of 824 m^2 with an OMS. This range was obtained by using 24 cameras (Spörri, Schiefermüller, & Müller, 2016). However, there are some limitations found in practical use, such as cost, calibration, portability, synchronization and set-up labor.

In addition, electromagnetic measurement systems (EMSs), image processing systems (IMSs), ultrasonic measurements (UMSs), and inertial measurement units (IMUs) are also applied in motion capture analyses in sports and physical exercise. EMSs can capture large amounts of information, but their measurement accuracy falls short of OMSs (Schepers & Veltink, 2010). IMSs area marker-less MCA system to digitally analyze captured photos and films (Zhong & Chang, 2004; Liuna, Santiago, Defez, Dunai, & Peris-Fajames, 2011). UMSs are generally used for short-range measurements. However, in a dynamic environment, controlling the directions of ultrasound waves could be a problem which reduces its practical applications (Bischoff,

Heidmann, Bust, & Paul, 2012). IMUs measure, process, and account for specific forces of a body, rate of angular movement, and the orientation of the object. They are attached to body segments to collect linear, rotational and angular data to establish the orientations of the body and detect high speed movements. In gymnastics, swimming and other sports, IMUs are most commonly used for MCA. However, they are usually sensitive to magnetic materials in the study environment (Lee, Burkett, Thiel, & James, 2011).

In combining motion capture analyses with other objective and subjective assessments, researchers have examined different types of activewear and accessories in dynamic wear. For example, Liu (2015) studied the biomechanical efficacy of knee bracing on knee kinetics and kinematics by performing a motion capture analysis with a Vicon motion capture system (Oxford Metrics, UK), a 3D gait analysis, balance test, and subjective survey (Fig. 5.17A). In this study, the knee joint related movements, angles, ground reaction forces in three anatomic planes, and temporal-spatial variables were examined. Wang, Wang, Kuo, and Su (2017) applied optical tracking system to capture the motion of the nude breast and breasts in bras at different levels of support during physical activities (Fig. 5.17B). By using these technologies, the biomechanical characteristics and user comfort of functional clothing can be analyzed, which provide an invaluable basis for new design and development of functional clothing and assistant units with improved mechanical performance for end users.

With advances in low-power circuitry and sensor technology, motion-tracking smart garments for athletes provide a complete picture of body motion and even give real-time feedback in normal sporting environments. These clothing-based MCA systems provide quantitative data that allow athletes to carry out repeatable and consistent movements more effectively (StretchSense, 2018). A new cameraless motion monitor has been reported which introduces a very thin sensor module fabricated by using hetero-core optical fibers, which is then inserted into sportswear to detect elbow, joint and trunk movement in real time. This system is likened to an unconstrained motion capture system and originally intended to be used as a teaching device in the sports and rehabilitation fields (Koyama et al., 2013). Garment-based motion capture (GaMoCap) which uses a garment that has printed color-coded patterns, has been reported. The coded patterns allow a high-density distribution of markers per unit of the surface area (approx. 40 markers per 100 cm^2) to realize the simultaneous reconstruction of shape

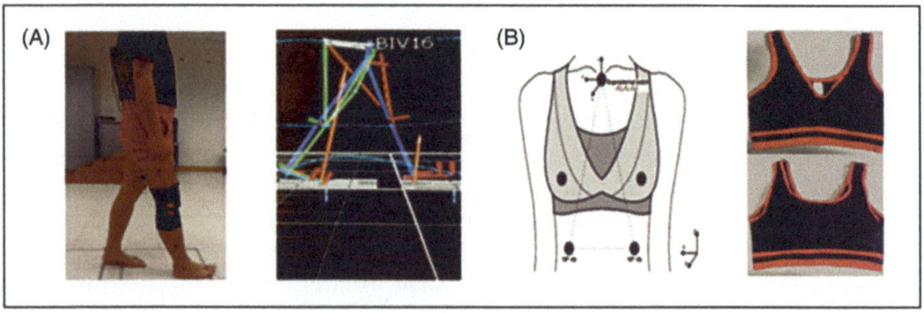

Figure 5.17 Motion capture analysis of (A) knee brace and (B) bra.

and motion. By assessing three different articulated motions (shoulder, knee, and hip flexion-extensions), Biasi et al. (2015) reported that the GaMoCap has a comparable accuracy as a commercial VICON system.

With the development of advanced algorithms, computing and informatics technologies, numerical simulation methods are increasingly being applied in modeling, predicting, and visualizing the functional performance of functional clothing, including activewear. The numerical approach not only facilitates researchers to further understand complex processes that are not viable by using conventional experiments but can also intuitively and visually show results that are affected by a multitude of factors without actually using materials or fabricating garments. For example, Wang, Luo, Li, Dai, and You (2005) used a geometric interpolatory volumetric subdivision scheme over the hexahedron lattice to simulate elastic deformation and pressure distribution of the human body with a tight-fit garment. Liu et al. (2006, 2007) developed finite element (FE) models to numerically simulate the lower body of women to evaluate effects of compression garments (e.g., compression stockings) with different physical–mechanical properties on skin and the underlying tissues during static and dynamic wear. Yu, Yick, Ng, Yip, and Chan (2016) developed a biomechanical model to simulate pressure profile of hand dorsum with a glove. Liu and Xu (2019) further developed 3D biomechanical model to analyse skin surface pressure and internal tissue stress variations by multi-class compression garment. Sun et al. (2019) also developed an FE model but applied a sensitivity analysis on bra design features to predict the effects of bra deformation on comfort pressure and breast shape.

5.6 Conclusion

Activewear is a type of functional garment that is being increasingly used in sporting activities or exercise to facilitate optimal physical performance and body motion. This type of garment also increases the wear comfort of users regardless of the climatic or environmental conditions. The physical, mechanical, chemical, and biomechanical properties of fabrics influence activewear and its functional performance and perceived comfort in the complex interactions of clothing, wearer, environment, and exercise. Evaluations of the physical, psychological, physiological, and biomechanical performances of activewear not only help to provide a better understanding on design effectiveness of activewear and related products for practical use, but also contribute to improving and optimizing design solutions that best meet the overall needs of the target consumers.

Acknowledgements

We sincerely thank the General Research Fund project (PolyU 252153/18E) supported by Research Grant Council of the Government of the Hong Kong Special Administrative Region, research project (ITP/097/18TP) funded by Innovation and Technology Fund, and Postgraduate Research Program (88011) and Department General Research project (UAHB) funded by the Hong Kong Polytechnic University to support this study.

References

American Association of Textile Chemists and Colorists (2019). AATCC. Retrieved from: https://www.aatcc.org/.

American Association of Textile Chemists and Colorists (2014). AATCC TM81.pH of the Water-Extract from Bleached Textiles. Retrieved from: https://members.aatcc.org/store/tm81/500/.

Ashdown, S. (2017). Modeling of half-scale human bodies in active body positions for apparel design and testing. *International Journal of Clothing Science and Technology*, *29*(6), 807–821.

ASTM (2019). ASTM International - Standards Worldwide. Retrieved from: https://www.astm.org.

Baddeley, A. (1992). Working memory: The interface between memory and cognition. *Journal of Cognitive Neuroscience*, *4*(3), 281–288.

Beaudette, E., & Park, H. (2017). Impact of seam types on thermal properties of athletic bodywear. *Textile Research Journal*, *87*(9), 1052–1059.

Bell, R., Cardello, A., & Schutz, H. (2005). Relationship between perceived clothing comfort and exam performance. *Family and Consumer Sciences Research Journal*, *33*(4), 308–320.

Biasi, N., Setti, F., Bue, A. D., Tavernini, M., Lunardelli, M., Fornaser, A., Da Lio, M., & Cecco, M. D. (2015). Garment-based motion capture (GaMoCap): higih-density capture of human shape in motion. *Machine Vision and Applications*doi: 10.1007/s00138-015-0701-2.

Bini, R. R., & Carpes, F. P. (2017). Segmental movements in cycling. In B. Müller et al.,(Ed.), *Handbook of human motion*. Cham: Springerdoi: 10.1007/978-3-319-30808-1_135-1.

Bischoff, O., Heidmann, N., Bust, J., & Paul, S. (2012). Design and implementation of an ultrasonic localization system for wireless sensor networks using angle-of-arrival and distance measurement. *Procedia Engineering*, *47*, 953–956.

BS (2019). British Standards. Retrieved from: https://shop.bsigroup.com/Navigate-by/Standards/.

DenHartog, E. A., & Deaton, A. S. (2015). Combined thermal manikin and thermal model predictions of working times in fully encapsulated impermeable suits. *Extreme Physiology & Medicine*, *Vol. 4*(Suppl 1), ppA142.

Deng, Y. M., Wang, S. F., & Wang, S. J. (2016). Study on antibacterial and comfort performances of cotton fabric finished by chitosan-silver for intimate apparel. *Fibers and Polymers*, *17*(9), 1384–1390.

Choudhury, A. R., Majumdar, P. K., & Datta, C. (2011). Factors affecting comfort: Human physiology and the role of clothing. In *Improving comfort in clothing* (pp. 3–60). Woodhead Publishing.

Corazza, S., Mündermann, L., Gambaretto, E., & Ferrigno, G. (2010). Markerless motion capture through visual hull, articulated icp and subject specific model generation. *International Journal of Computer Vision*, *87*(1–2), 156.

Fan, Q. (2008). Fabric chemical testing-5. In *Fabric testing* (pp. 125–147). Elsevier Ltd.

Farnworth, B., & Dolhan, P. (1984). Apparatus to measure the water-vapour resistance of textiles. *Journal of the Textile Institute*, *75*(2), 142–145.

GB Standards (2019). What is GB standards. Retrieved from: http://www.gbstandards.org/.

Hes, L., & Williams, J. (2011). Laboratory measurement of thermo-physiological comfort-5. In *Improving comfort in clothing* (pp. 114–137). Elsevier Ltd 2011.

Huang, J. (2006). Sweating guarded hot plate test method. Polymer testing, 25(5), 709-716.

Iggo A. Sensory Receptors, Cutaneous, in *Sensory Systems*, I.I., Senses Other Than Vision, J.M., Wolye (Ed.), 1988: Basel, A Pro Scienta Viva Title, pp. 109–110.

International Organization for Standardization (ISO) (2001). Measurement of thermal insulation by means of a thermal manikin (Standard No. ISO/DIS 15831:December 2001). Geneva, Switzerland: ISO.

ISO (2019). International Organization for Standardization. Retrieved from: https://www.iso.org.

ISO 11092. (2014). Textiles—Physiological effects—Measurement of thermal and water-vapour resistance under steady-state conditions (sweating guarded-hotplate test). Standard test method.

JIS (2019). Japanese Industrial Standards. Retrieved from: https://www.jisc.go.jp/.

Hawley, J. (2000). Biomechanics of Running. *Handbook of sports medicine and science: running*. Oxford: Blackwell Science pp. 28-43.

Holmér, I. (2004). Thermal manikin history and applications. *European Journal of Applied Physiology*, *92*(6), 614–618.

Kamalha, E., Zeng, Y., Mwasiagi, J. I., & Kyatuheire, S. (2013). The comfort dimension; a review of perception in clothing. *Journal of sensory studies*, *28*(6), 423–444.

Kilinc-Balci, F. (2011). How consumers perceive comfort in apparel-4. In *Improving comfort in clothing* (pp. 97–113). (2011). Elsevier Ltd.

Kirschner, & Paul, A. (2002). Cognitive load theory: Implications of cognitive load theory on the design of learning. *Learning and Instruction*, *12*(no. 1), 1–10.

LaBat, K. L., & DeLong, M. R. (1990). Body cathexis and satisfaction with fit of apparel. *Clothing and Textiles Research Journal*, *8*(2), 43–48.

Lai, C. H. Y., & Li-Tsang, C. W. P. (2009). Validation of the pliance X system in measuring interface pressure generated by pressure garment. *Burns*, *35*(6), 845–851.

Lee, E. K., Hyun, K. J., & Tokura, H. (2000). The effect of skin pressure by clothing on cicardian rhythms of core temperature and salivary melatonin. *Chronobiology International*, *17*, 783–793.

Lee, J. B., Burkett, B. J., Thiel, D. V., & James, D. A. (2011). Inertial sensor, 3D and 2D assessment of stroke phases in freestyle swimming. *Procedia Engineering*, *13*, 148–153.

Lexico (2019). In Oxford Dictionary. Retrieved from: https://en.oxforddictionaries.com/definition/activewear.

Likert R (1932). A Technique for the Measurement of Attitudes. Archives of Psychology; No. 140. Dayton, Ohio: NCR. New York University. Bibliography, pp. 54–55.

Liu R (2015). *Biomechanical efficacy of knee bracing on joint kinetics and kinematics, postural steadiness and wearing perception*. The Hong Kong Polytechnic University.

Liu, R., & Little, T. J. (2009). Optimizing 5Ps for compression athletic wear comfort in sports. *Journal of Fiber Bioengineering and Informatics*, *2*(1), 44–55.

Liu, R., Kwok, Y. L., Li, Y., Lao, T. T., Zhang, X., & Dai, X. Q. (2005). Objective evaluation of skin pressure distribution of graduated elastic compression stockings. *Dermatologic Surgery*, *31*(6), 615–624.

Liu, R., Kwok, Y. L., Li, Y., Lao, T. T., & Zhang, X. (2007). Skin pressure profiles and variations with body postural changes beneath medical elastic compression stockings. *International Journal of Dermatology*, *46*(5), 514–523.

Liu, R., Little, T. J., & William, J. R. (2014). Compression form-fitted athletic wear: pressure performance, moisture management properties under different tension ratios, and corresponding psychophysical responses. *Fibers and Polymers*, *15*(3), 632–644.

Liu, R., Kwok, Y. L., Li, Y., Lao, T. T., & Zhang, X. (2006). A three-dimensional biomechanical model for numerical simulation of pressure functional performances of graduated compression stocking (GCS). *Fibers and Polymers*, *7*(4), 389–397.

Liu, R., Kwok, Y. L., Li, Y., Lao, T. T., Dai, X. Q., & Zhang, X. (2007). Numerical simulation of inner stress profiles and deformations of lower extremity beneath graduated compression stockings. *Fibers and Polymers*, *8*(3), 301–308.

Liu, R., Kwok, Y. L., & Lao, T. T. (2012). Effects of skin pressures by compression legwear on resting salivary cortisol and urinary catecholamines excretion in women. *Dermatologic Surgery*, *38*(1), 83–90.

Liu, R., Lao, T. T., Kwok, Y. L., Li, Y., & Ying, M. (2008). Effects of graduated compression stockings with different pressure profiles on lower-limb venous structures and haemodynamics. *Advances in Therapy*, *25*(5), 465–478.

Liu, R., Little, T., & Williams, J. R. (2012). Evaluation of elite athletes psycho-physiological responses to compression form-fitted athletic wear in intensive exercise based on 5Ps model. *Fibers and Polymers*, *13*(3), 380–389.

Liuna, E., Santiago, V., Defez, B., Dunai, L., & Peris-Fajames, G. (2011). Velocity vector (3D) measurement for spherical objects using an electro-optical device. *Measurement*, *44*(9), 1723–1729.

Maletsky, L. P., Sun, J., & Morton, N. A. (2007). Accuracy of an optical active-marker system to track the relative motion of rigid bodies. *Journal of Biomechanics*, *40*(3), 682–685.

Marqués, L., Cela, J., & Gisbert, M. (2017). Pre-service physical education teachers' self-management ability/: a training experience in 3D simulation environments. *Retos Nuevas Tendencias En Education Fisia, Deport Y Recreación*, *32*, 30–34.

Milenković, L., Škundrić, P., Sokolović, R., & Nikolić, T. (1999). Comfort properties of defence protective clothing. *The Scientific Journal Facta Universities*, *1*(4), 101–106.

Millet, G., Perrey, S., Divert, C., & Foissac, M. (2006). The role of engineering in fatigue reduction during human locomotion—a review. *Sports Engineering*, *9*(4), 209–220.

Miyasuji, A., Matsumoto, T., Mitarai Sk, Kotabe, T., Takeshima, T., & Watanuki, S. (2002). Effects of clothing pressure caused by different types of brassieres on automic nervous system activity evaluated by heart rate variability power spectral analysis. *Journal of Physiological Anthropology*, *21*, 67–74.

Naugle, L. M. (1999). Motion capture: Re-collecting the dance. *Proc. ICKL*, *99*, 208–213.

Reed, C. C., Wolf, W. A., Cotton, C. C., & Dellon, E. S. (2017). A visual analogue scale and a Likert scale are simple and responsive tools for assessing dysphagia in eosinophilic oesophagitis. *Alimentary Pharmacology & Therapeutics*, *45*(11), 1443–1448.

Schepers, H. M., & Veltink, P. H. (2010). Stochastic magnetic measurement model for relative position and orientation estimation. *Measurement Science and Technology*, *21*(6), 065801.

Senthilkumar, M., & Anbumani, N. (2011). Dynamics of elastic knitted fabrics for sportswear. *Journal of Industrial Textiles*, *41*(1), 13–24.

Sesselmann, G.J. (1996). U.S. Patent No. 5,539,930. Washington, DC: U.S. Patent and Trademark Office.

Sesselmann, G.J. (2000). U.S. Patent No. 6,134,718. Washington, DC: U.S. Patent and Trademark Office.

Shi, H., Zhang, L., & Wang, J. (2017). Influence of seam types on seam quality of outdoor clothing. *International Journal of Clothing Science and Technology*, *29*(4), 553–565.

Shishoo, R. (2005). *Elastic Textiles Textiles in Sport*. Woodhead Publishing Limited, Woodhead Publishing.

Spörri, J., Schiefermüller, C., & Müller, E. (2016). Collecting kinematic data on a ski track with optoelectronic stereophotogrammetry: A methodological study assessing the feasibility of bringing the biomechanics lab to the field. *PLOS One*, *11*(8), e0161757.

StretchSense (2018). Motion capture clothing-the new age of wearable tech. Retrieved from: https://stretchsense.com/article-resources/blog/motion-capture-clothing-the-new-age-of-wearable-tech/.

Suganthi, & Senthilkumar (2018). Moisture-management properties of bi-layer knitted fabrics for sportswear. *Journal of Industrial Textiles*, *47*(7), 1447–1463.
Sun, Y., Yick, K. L., Yu, W., Chen, L. H., Lau, N., Jiao, W. Z., & Zhang SC (2019). 3D bra and human interactive modeling using finite element method for bra design. *Computer-aided Design*, *114*, 13–27.
Tessier, D. (2018). Testing thermal properties of textiles. In Advanced characterization and testing of textiles (pp. 71-92). Woodhead Publishing.
Testex textile. (March 16, 2020). ISO 11092 - 2014 Guide: testing of thermal resistance and water vapour resistance of textile fabrics. Retrieved from https://www.testextextile.com/iso-11092-2014-guide-testing-of-thermal-resistance-and-water-vapour-resistance-of-textile-fabrics/.
Tyler, D. (2018). *Application of pressure sensors in monitoring pressure, in materials and technology for sportswear and performance apparel.* CRC Press pp. 300-321.
Wang Dong (2018). Functional sportswear fabric not easy to produce bacteria or mould. Retrieved from: https://worldwide.espacenet.com/publicationDetails/biblio?FT=D&date=20180306&DB=EPODOC&locale=&CC=CN&NR=107756899A&KC=A&ND=1#.
Wang, C. S., Wang, L. H., Kuo, L. C., & Su, F. C. (2017). Comparison of breast motion at different levels of support during physical activity. *Journal of Human Sports and Exercise*, *12*(4), 1256–1264.
Wang, J. M., Luo, X. N., Li, Y., Dai, X. Q., & You, F. (2005). The application of the volumetric subdivision scheme in the simulation of elastic human body deformation and garment pressure. *Textile Research Journal*, *75*, 591–597.
Wang, Y. R., Cui, Y. H., Zhang, P. H., & Feng, X. W. (2011). A smart mannequin system for the pressure performance evaluation of compression garments. *Textile Research Journal*, *81*(11), 1113–1123.
Yu, A., Yick, K. L., Ng, S. P., Yip, J., & Chan, Y. F. (2016). Numerical simulation of pressure therapy glove by using Finite Element Method. *Burns*, *42*(1), 141–151.
Yu, W. N., Fan, J. T., Qiao, X. M., & Tao, X. M. (2004). A soft mannequin for the evaluation of pressure garments on human body. *Sen-I Gakkasishi*, *60*(2), 57–64.
Zhao, R.Z., & Wang, J.G. (2013). Thin light windproof warm-keeping humidity-conducting anti-bacteria outdoor sportswear fabric. Patent, CN103110222 (A) -2013-05-22.
Zhong, D., & Chang, S. F. (2004). Real-time view recognition and event detection for sports video. *Journal of Visual Communication and Image Representation*, *15*(3), 330–347.

Further reading

Brophy-Williams, N., Driller, M. W., Halson, S., & Fell, J. (2014). Evaluating the Kikuhime pressure monitor for use with sports compression clothing. *Sports Engineering*, *17*(1), 55–60.
CEN (2019). European Committee for Standardization. Retrieved from: https://www.cen.eu/Pages/default.aspx.
Garments Tech (2019). Garment pH value & its test procedure. Retrieved from: http://garmentstech.com/garment-ph-value-test-procedure/.
Koyama, Y., Nishiyama, M., & Watanabe, K. (2013). A motion monitor using hetero-core optical fiber sensors sewed in sportswear to trace trunk motion. *IEEE Transactions on Instrumentation and Measurement*, *62*(4), 828–836.
Liu R, Xu B. (2019). 3D digital modeling and design of custom-fit functional compression garment. In W.K. Wong, (ed.) *Artificial Intelligence on Fashion and Textiles, Advances in Intelligent Systems and Computing*, 849, https://doi.org/10.1007/978-3-319-00695-0_20.

Liu, S., & Zhou, J. (2019). Analysis of the dynamic properties of elastic knitted fabric for sportswear: Inverse stress relaxation. *Textile Research Journal*, *89*(9), 1673–1683.

Manufacturing Solutions Center (2019). Retrieved from: https://www.manufacturingsolutionscenter.org/moisture-management-testing.html.

Moeslund, T. B. (2000). Interacting with a virtual world through motion capture. In *Virtual Interaction: Interaction in Virtual Inhabited 3D Worlds* (pp. 221–234). Springer.

Paul, R. (2019). High performance technical textiles: An overview. In *High Performance Technical Textiles* (pp. 1–10). Chichester, UK: John Wiley & Sons.

SGS (2011). Measuring and explaining the "Creature Comforts" of Performance Textiles. Retrieved in July 2019 from https://www.aatcc.org/wp content/uploads/2014/11/ORWinterMarketFunctionalTestingJan202011.ppt.

Standards Association of Australia (1989). AS 2001.2.15:1989. Methods of test for textiles – physical tests – determination of thickness of textile fabrics. *Australian Standards*.

Standards Association of Australia (1991). AS 2001.2.5:1991. Methods of test for textiles – physical tests – determination of the number of threads per unit length in woven fabric. *Australian Standards*.

Standards Association of Australia (2001). AS2001.2.6:2001. Method of test for textile – physical test - determination of the number of wales and courses per unit length in knitted fabric. *Australian Standards*.

Standards Australia (2019). Australian Standards. Retrieved in July 2019 from https://www.standards.org.au/.

Troynikov, O., Nawaz, N., & Watson, C. (2018). Durability of vapor-permeable waterproof textile materials used in sailing protective apparel. *Textile Research Journal*, *88*(24), 2825–2840.

Wu, X. B., & Liu, R. (2020). Modern textile-based compression device for improving venous haemodynamics of lower extremities. *Advances in Physical Ergonomics and Human Factors*. doi: 10.1007/978-3-030-20142-5_44.

Compression technology

6

Kit-lun Yick
Institute of Textiles and Clothing, The Hong Kong Polytechnic University, Hung Hom, Hong Kong

6.1 Introduction

Today, activewear and sportswear are becoming an important sector of the global fashion industry. With increased demand for innovative designs and the growing awareness of maintaining healthy and active lifestyles, the global sports apparel market is expected to increase from US$173.68 billion to a projected US$219.76 billion by 2024 (Statista, 2019). Amongst the different types of activewear, there is one that is highly technical as well as functional and enhances performance, which is called performance sportswear. Performance sportswear is very expensive and produced in small volumes (Manshahia & Das, 2014).

More recently, the principles of compression garments are being applied to performance sportswear for medical purposes. Compression textiles, which are fabricated from elastomeric fibers and yarns for high stretch and elastic recovery, are generally used to not only improve garment fit and range of body movement, but also strategically apply an appropriate amount of mechanical pressure onto the body to support and stabilize the muscles. The early uses of compression garments were for the clinically treating and rehabilitation of hypertrophic scars, treating chronic venous diseases, providing orthopedic support, etc.

Hypertrophic scarring occurs after a wound heals due to the excessive deposition of wound collagen. Pressure therapy is usually used as a form of treatment, and the rationale for doing so is that pressure reduces collagen production within a developing or an active scar (Macintyre & Baird, 2006). Compression garments must fit like a second skin and are engineered to apply adequate pressure onto various body parts with the aim to increase the rate of scar maturation, prevent contracture formation, and enhance cosmetic appearance without impairing circulation. For optimal healing of a hypertrophic scar, compression garments must be used for 12–18 months or until the scar matures. To treat chronic venous insufficiency, extra firm compression stockings that provide a high range of pressure onto the ankles, 0.30–0.38 kPa (40–50 mmHg), might be necessary to reduce occupational leg swelling (Liu, Kwok, Li, Lao, Zhang & Dai, 2006; Liu, Kwok, Li, Lao, & Zhang, 2007). Knitted orthopedic supports such as knee sleeves for degenerative knee joints and osteoarthritis are intended to strengthen the quadriceps muscle group including the rectus femoris, and vastus lateralis, medialis, and intermedius muscles (Liu, Guo, Lao, & Little, 2017). Additional metal or silicone inserts could also be embedded for reinforcement, thereby increasing the mobility of patients during daily activities.

The design and development of performance sportswear widely vary in style, fabrication complexity and cost, and depending on their end-uses, the level of protection,

Latest Material and Technological Developments for Activewear. http://dx.doi.org/10.1016/B978-0-12-819492-8.00006-5

fit, comfort, etc. They are generally form-fitted and custom-engineered for specific athletes and competitions to enhance performance, reduce injuries, promote recovery, and enhance overall circulation with a faster warm up time. It is anticipated that the feelings of support provided by compression activewear could enhance augmented proprioception for improved muscle coordination, whilst less muscle vibration and oscillation could improve recovery with more rapid muscle repair, and minimize swelling following exercise and training sessions (Manshahia & Das, 2014). Nevertheless, the effects of compression activewear on exercise performance are somewhat inconclusive. Only a few studies are available that conclude on the positive effects of compression activewear on exercise performance, submaximal and maximal endurance performance (arterial lactate concentration, oxygen saturation, and oxygen uptake) and muscle metabolism during recovery (muscle blood flow and glucose uptake) (MacRae, Cotter, & Laing, 2011; Mizuno, Arai, Todoko, Yamada, & Goto, 2017; Morooka, Nakahashi, Morroka, & Kitamura, 2001; Sperlich, Born, Kaskinoro, Kalliokoski, & Laaksonen, 2013). The inconclusiveness of previous studies is also attributed to the differences in experimental conditions, design, and levels of pressure provided by compression activewear, as well as the subject variations (such as differences in their body weight and size, anatomy, fat proportion, motion, etc.) (Mizuno et al., 2017).

Designing compression activewear that exerts adequate pressure onto different body parts to effectively support the targeted muscles for better performance and quicker recovery is a challenging task. The pressure applied onto the body primarily depends on the mechanical properties of the garment which are very much based on the nonlinear stress–strain properties of the textile materials. Knitted fabrics fabricated from elastomeric fibers and yarns that demonstrate outstanding extensibility and elastic recovery are commonly used. With advances in knitting technologies, elastomeric yarns can be inlayed, floated, or plated into a knitted structure (Xiong & Tao, 2018). The compression performance of a fabric can be managed by controlling the density, thickness, and stiffness of the inlay yarns during the knitting process; see Fig. 6.1.

Today, one of the most recent and important applications for designing and engineering sports and performance garment designs is body mapping (Shishoo, 2015). This application takes into consideration how different areas of the body respond to the level of activity and the type of environment. Increased surface roughness

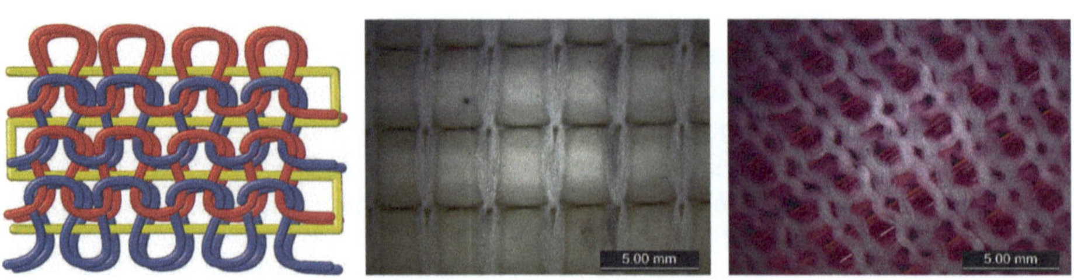

Figure 6.1 Laid-in knitted structure.

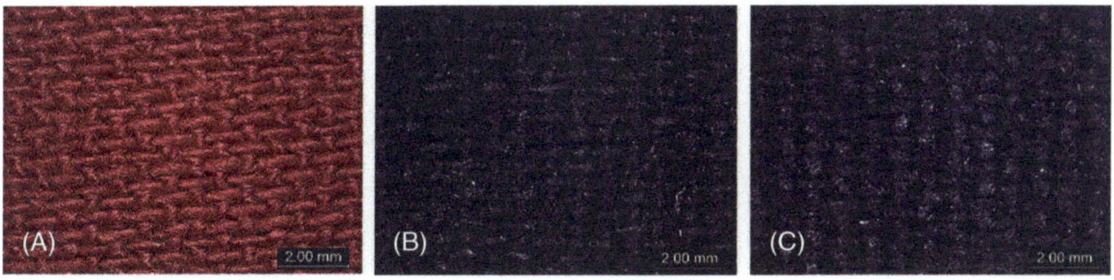

Figure 6.2 Fabric structure of different kinesiology tape: (A) Kinesio Tex Classic, (B) ATex, and (C) 3M Futuro.

on a sports garment improves aerodynamic performance in high-speed sports, and the hydrodynamic properties of swimwear fabric are typically and strategically used in designing compression activewear. Power bands that are used for sports taping or multiple reinforcement layers and/or inserts made of rigid materials may also be incorporated into compression activewear to control the range of motion, increase stability and reduce oscillation of the targeted muscles based on the anatomical curves and exoskeletal system. Sports tape, such as kinesiology tape, can offer therapeutic benefits by imposing tensile force onto the joints and muscles for clinical treatment and treating sports related injuries; see Fig. 6.2. The design of sports tape sets out to mimic the superficial layers of the human skin, or the epidermis, in terms of its capability to stretch (Lemos, Pereira, Protássio, Lucas, & Matheur, 2015; Vered, Oved, Ziberg, & Kalichman, 2016) and its thinness. Thus, the tape is not readily felt by the wearer but proprioceptive input; that is, the sensations from muscles and joints, to the brain and body are facilitated. In addition, kinesiology tape conforms to the body with the ability to stretch up to 40%–60% of its resting length, which is similar to the elastic properties of skin (Halseth, McChesney, DeBeliso, Vaughn, & Lien, 2004; Lemos et al., 2015) and thus allows for normal mobility and a complete range of motion (Yasukawa, Patel, & Sisung, 2006). Depending on the elastic quality, kinesiology tape could theoretically facilitate or inhibit muscle activity, promote good circulation and edema reduction, support joint structures and improve proprioception and stability (Huang, Hsieh, Lu, & Su, 2011; Vered et al., 2016).

More recently, body-mapped compression sportswear can also be produced by using seamless knitting technology to provide anatomic compression on the body of athletes (Shishoo, 2015). Materials with different rigidity and strength can be deliberately mapped to the targeted muscle for enhanced support, cushioning, moisture wicking, breathing, etc.

However, compression garments are usually fabricated by using the conventional cut-and-sew method, based on two-dimensional pattern pieces. To create compression, the circumferential dimensions are reduced by 10%–30% on the pattern in accordance with the fabric extensibility. Hence, size together with the reduction factor in pattern designs should be modified based on the type of materials used to produce the overall intended amount of compression or used as an additional layer of fabric that is engineered to provide compression in strategic areas for muscle support.

6.2 Current design of compression activewear

Activewear that has a moderate distribution of compression are popular garments for athletic and fitness activities. These types of garments are designed to enhance the performance of the athletes, reduce the possibility of injuries, and accelerate the process of recovery without limiting body movements. In comparison to pressure therapy garments, the design of compression activewear is highly complex and uses different types of fabric and different design features to not only address the requirements for both function and aesthetics, but also exemplify an active lifestyle. Moreover, the garments are engineer designed to fit the contours of the body and strengthen the outer layer to protect the joints and key muscles. To protect the wearer from the mechanical stress caused by various athletic activities, garment manufacturers have developed compression activewear with specific features such as those that increase muscle support, connect key muscle groups for greater power return, incorporate ergonomic seams for greater range of body movement, have increased reinforcement for cushioning and protection, etc. The material, environmental conditions, nature of the sport, tactile and ergonomic comfort, level of physical activity, and the specific requirements of the wearer need to be taken into consideration so that designing suitable compression activewear becomes a challenging task. Some of the novel designs of commercially available compression activewear are discussed below, including cross training and running compression tights, training and sports tights, cycling tights, as well as compression swimwear.

6.2.1 Cross-training compression tights

Cross training involves a range of dynamic movements that are physically demanding. Cross-training compression tights are generally made of high gauge elastomeric yarns and utilize a circular knit structure for stretch which would deliver flexibility and power and allow durability. 2XU, a sportswear company, claims that their compression tights which have a pressure range between 0.19 and 0.23 kPa (25–30 mmHg) can support the key muscles of the lower limbs including the glutes, hamstrings, quads, and calves which are subjected to high intensity impact from running, jumping, and squatting.

As for thermal comfort, materials that can dissipate heat and wick moisture away from the skin surface in anticipation of high relative humidity are used, which not only keep the wearer dry and cool for optimal comfort, but also prevent the growth of bacteria and fungi which cause odor. To enhance fit and comfort, panels such as those for the crotch can be pre-shaped and/or contour molded, whilst flatlock seams or even seamless adhesive bonding is used to reduce the seam profile and chafing. Since the amount of pressure prescribed by any compression garment must be carefully controlled, the sizing system of cross-training compression tights is formulated based on the body height and weight of the wearer. For example, the size of 2XU garments ranges from XXS (i.e., a body height of 145–165 cm and weight of 30–40 kg) to XXL (155–190 cm and 95–110 kg).

To facilitate the circulation of blood in the legs during prolonged and intensive training, compression tights need to provide graduated compression with high levels of compression at the ankles (e.g., 100%) and a gradual reduction up the leg

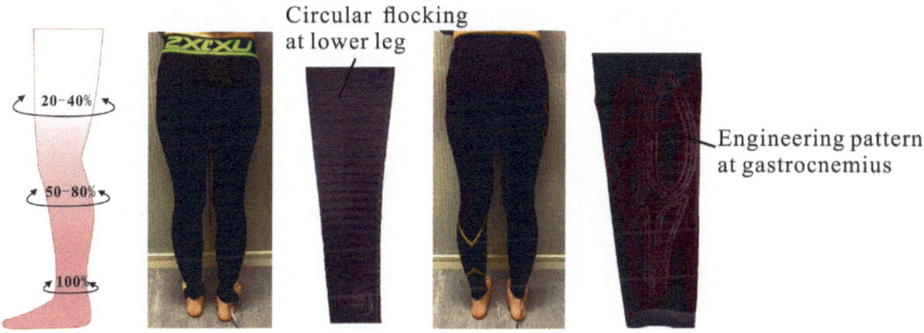

Figure 6.3 Criss-cross and/or flocking pattern specifically placed in targeted area (the calf) of compression to control muscles.
Source: 2XU (2019).

(20%–40%); see Fig. 6.3. That is, the compression level increases progressively from the thighs to the ankles by increasing the reduction factor of the pattern and/or the use of rigid reinforcement material with tighter elasticity so that the shape of the leggings is maintained. Criss-cross or flocking patterns, rigid adhesive, and/or silicone tape are, therefore, strategically used and/or applied, particularly at the ankles. As a result, the compression tights provide anatomical mapping for targeted control of the amount of pressure and support which would reduce muscle strain from lifting and squatting.

6.2.2 Running compression tights

Runner's knee is used to depict any injury to the knees due to overuse and repetitive impacts of the foot to the ground, thus resulting in wearing of the cartilage in the kneecap, and subsequently knee pain. Achilles tendinitis may also occur as a result of overuse, which leads to pain and stiffness in the tendon area and tight calf muscles.

In response, a high performance sports apparel company called CW-X patented the EXO-WEB technology which mimics kinesiology-taping techniques for sports injuries and sore joints (CW-X, 2019). In doing so, the EXO-WEB technology creates a support system for the exoskeleton which enhances biomechanical performance when the user is moving. The CW-X EXO-WEB technology allows two-way stretching, which provides targeted support to key areas such as the vastus lateralis, biceps femoris, fibularis longus, and patella muscles, and, therefore, reduces muscle oscillation. By controlling the elasticity of the web, the variable compression along the web facilitates the circulation of blood and increases oxygen intake, thus minimizing the build-up of lactic acid and muscle soreness for a shorter recovery time and improved athletic performance (Fig. 6.4).

In contrast, thermal compression tights have also been designed to improve insulation against the cold for outdoor running. The tights not only provide compression to help reduce muscle movement, but also improve blood flow for a faster warm up and recovery through graduated compression. Fabrics used for thermal compression apparel may be brushed high gauge knitted structures that improve the ability to maintain a still layer of air in the garment for thermal insulation purposes.

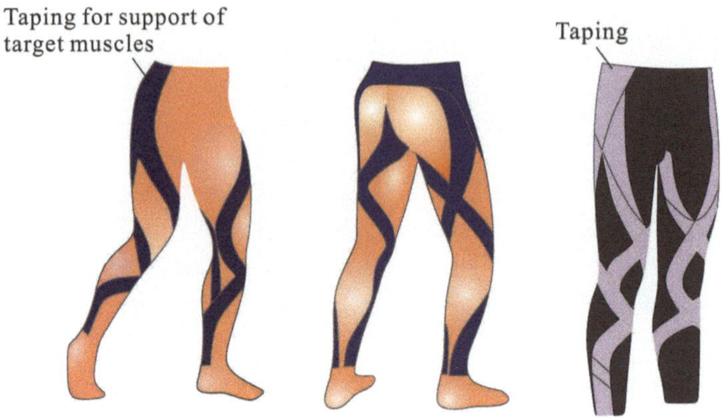

Figure 6.4 CW-X EXO-WEB technology creates support system for exoskeleton and contours lower limb muscles.
Source: CW-X (2019).

6.2.3 Training and sports tights

Training and sports tights which have relatively lower compression are used to shape the body. Their tighter fit is like a second layer of skin that gives the body a supporting feel. The amount of pressure exerted by these tights is low (<75 Pa or 10 mmHg). There is minimum support of key muscles.

As dynamic movement of the body produces heat and sweat, the capability of fabric to regulate temperature and humidity has a decisive influence on the thermal comfort of a garment. Heat loss is due to convection and evaporation through the pores of the material and openings. With advances in knitting technologies, mesh inserts and three-dimensional (3D) knitted mesh spacer fabrics are increasingly being adopted in compression activewear, as they offer superior cushioning; see Fig. 6.5. Their porous structure also enhances the breathability of the activewear (Liu, Hu, Zhao, & Long, 2012; Rajan, Souza, Ramakrishnan, & Zakriya, 2016). More recently, perforated panels with small-holes in specific patterns are frequently applied by using laser cutting technology thus allowing efficient release of heat from the body even though its production process is more costly. Because of the uneven distribution of sweat glands across the body, mesh and/or perforated panels are generally used on the back of the legs or at the outer thighs and lower legs to offer superior breathability and heat dissipation.

6.2.4 Cycling compression tights

Cycling compression shorts are made of lightweight and strong compression fabrics engineered for sports. The fabric can stretch in different directions for optimum comfort because this allows the wearer to freely move about. Compression shorts offer excellent stabilization of the muscles, and causes less muscle vibration and fatigue, while their moisture management properties keep the wearer dry and comfortable. In consideration of body movement during cycling, additional padding and/or material inserts (such as STEEL X fabric designed by 2XU which consists of polyethylene

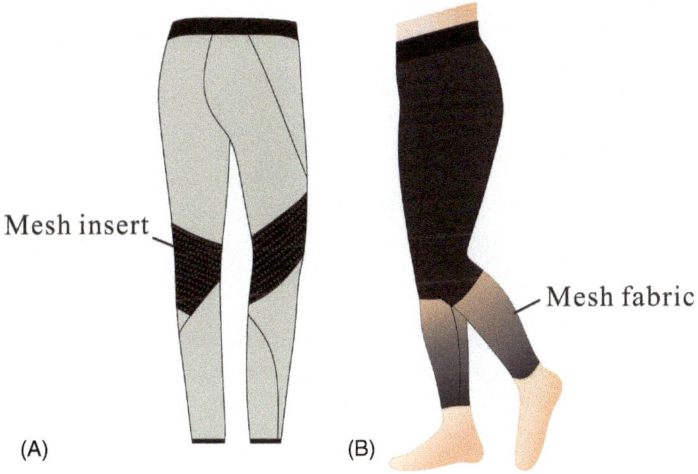

Figure 6.5 Mesh panels that enhance breathability: (A) Adidas Alphaskin Sport Tights and (B) NIKE Pro Hypercool.
Sources: Adidas (2019), NIKE (2019).

fibers that have an ultra-high molecular weight) are placed as hip panels for reinforcement purposes which offer ultimate protection to avoid abrasion from the bike and are cut and tear resistant. Embedded paddings and/or inserts are generally pre-shaped to ensure an ideal fit; see Fig. 6.6. Perforated patterns may also be used to enhance breathability and evaporation of moisture, thus ensuring wear comfort.

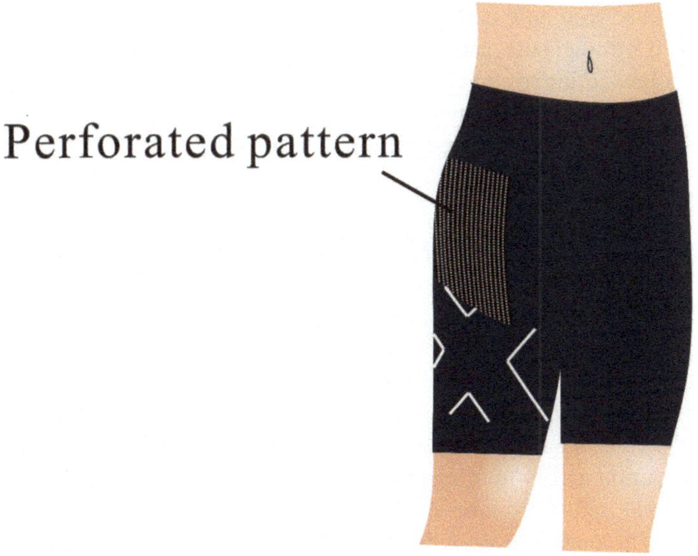

Figure 6.6 Cycling compression shorts with STEELX fabric from 2XU.
Source: 2XU (2019).

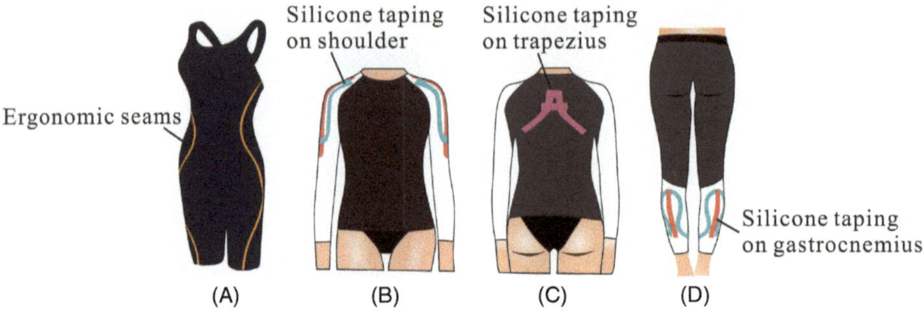

Figure 6.7 (A) SPEEDO Fastskin LZR Intent Openback, and ARENA women's carbon compression activewear (B) front, (C) back, and (D) tights.
Source: SPEEDO (2019), Arena (2019).

6.2.5 Compression activewear for swimming

Shishoo (2015) indicated that layering is the functional concept behind compression activewear and assembly, which influences body movement and maintains areas of thermos-neutrality. An example of compression activewear which is fabricated with multiple layers of fabric and ergonomic seams is the SPEEDO Fastskin LZR racers (swimwear) which compress and connect the key muscles, including the lateral dorsi and gluteus maximus muscles and hamstrings (Fig. 6.7(A)). Textured composite fabric is custom-engineered to reduce drag against water for optimal performance. Compression activewear may also incorporate carbon technology for intelligent and therapeutic compression of the muscles which would enhance circulation; for instance, ARENA women's carbon compression activewear (see Fig. 6.7(B)–(D)). Silicone tape is strategically used to prevent shoulder, trapezius and/or calf muscle injuries and allow faster recovery. Highly knitted fabrics are used for sculpting and shaping the body which would give the activewear a second-skin fit. Laser-cut holes in designated body areas are incorporated for optimal breathability and comfort.

6.3 Pressure induced by compression activewear

Amongst the different types of compression activewear, the pressure performance including the amount and duration of pressure is one of the most important criteria for evaluating the effectiveness of compression garments, but has provided inconsistent results. The literature on compression activewear shows that a compression pressure less than 0.06 kPa (7.5 mmHg) exerted on the trunk by a swimsuit could be helpful for venous pump action and swimming performance. Nevertheless, excessive pressure on the shoulders and inguinal region suppress blood flow to peripheral parts and causes delayed blood pressure recovery after exercise (Morooka et al., 2001). Pressure values that exceed 1.15 kPa on the chest, 0.36 kPa on the waist, and/or 0.56 kPa on the abdomen; however, may adversely affect the wear comfort of swimsuits. In the case of sports vests, the recommended pressure range for a comfortable fit is 0.96–1.35 kPa (Liu, Chen, Wei, & Pen, 2011). Nevertheless, the amount of pressure induced by compression activewear remains ambiguous. A pilot study has been, therefore, carried out below to investigate the effects of garment pressure distribution from different types of compression tights on young female subjects.

6.3.1 Experimental

A total of 10 young female subjects (who are between 22 and 28 years old) took part in a study conducted by a team at the Institute of Textiles and Clothing of the Hong Kong Polytechnic University. Their body mass index (BMI) ranges from 18.69 to 23.04 (with a mean BMI of 20.29, SD 1.17). According to size specifications, all of the subjects are a small size. A total of four different compression tights were sourced commercially. Details of the four types of leggings are presented in Table 6.1. Each participant donned the four different types of compression tights for four experimental conditions (Samples I, II, III, and IV in Fig. 6.8), in a randomized sequence. The participants were instructed to perform three body movements; namely, static standing, single leg standing, and standing with their heels upwards. Thus, there were a total of 12 experimental conditions which were carried out in random order. For each condition, the participants were tested thrice for 30 s and given a break of 1 min for resting between each condition. The garment pressure exerted on the right leg was measured on the upper part of the leg (rectus femoris muscle and hamstring) and lower part of the leg (gastrocnemius muscle), respectively.

Table 6.1 Details of compression tights

	Brand name	Product name	Features
Sample I	2XU	Elite power recovery	• Faster recovery and reduces swelling • Graduated compression with pressure range of 25–30 mmHg • Progressive inelastic criss-cross flocking from knee to hem to increase compression level at ankle
Sample II	2XU	MCS run compression tights	• Anatomical muscle mapping • Greater endurance and less muscle damage • Engineered flocking pattern placed at rectus femoris and gastrocnemius muscles
Sample III	2XU	Mid-rise compression tights	• Graduated compression with curved flatlock seams
Sample IV	Reecho	Performance pro tights	• Diagonal panel at front of thighs • Tape around knee caps for reinforcement

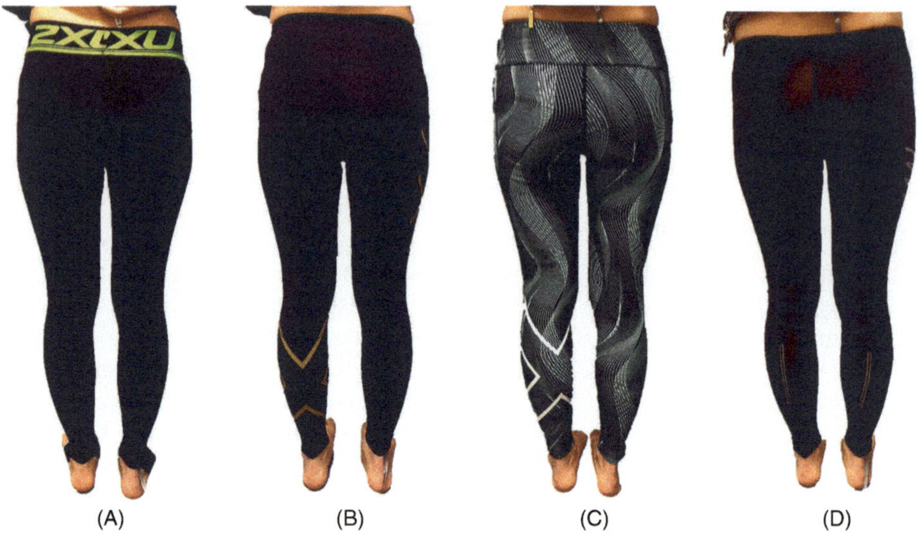

Figure 6.8 Four different types of compression tights: (A) Sample I, (B) Sample II, (C) Sample III, and (D) Sample IV.

The Novel Pliance-X pressure system is used in this study to measure the garment pressure. This device provides highly accurate pressure and sensor measurements. This pressure system also ensures good repeatability which contributes to accurate tights-skin pressure values for evaluating the different designs of compression tights.

6.3.2 Results of pressure evaluation

In the static standing condition, it is interesting that the pressure values obtained from the upper part of the legs are consistently higher than those from the lower part of the legs for all of the legging samples, regardless of their graduated compression. As shown in Fig. 6.9, Sample I has the highest mean pressure (3530 Pa) for all three pressure locations. With the exception of Sample II, the garment-skin pressure is evenly distributed at the rectus femoris muscle and the hamstrings (anterior and posterior of the upper part of the leg) for the other samples. The pressure obtained at the rectus femoris muscle (4829 Pa) for Sample II is substantially higher than that at the hamstrings (580 Pa), which could be attributed to the engineered pattern with a flocking effect on the compression tights at the rectus femoris muscle. The criss-cross and flocking effects at the gastrocnemius muscles of Samples I and II also lead to increase in pressure, as compared to Samples III and IV which do not have flocking and/or reinforcement inserts.

The effects of the seamlines on the pressure exerted on the lower part of the leg were also examined, with the exception of Sample I which does not have a leg seam; see Fig. 6.10. Pressure at lower part of leg along the seamline was measured and compared against the fabric region (without seam). It is worth noting that the flatlock

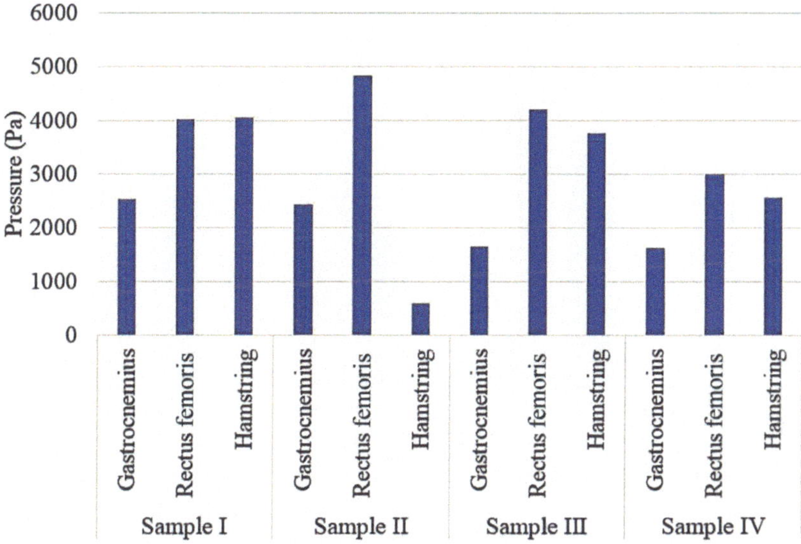

Figure 6.9 Pressure at different locations in static standing condition.

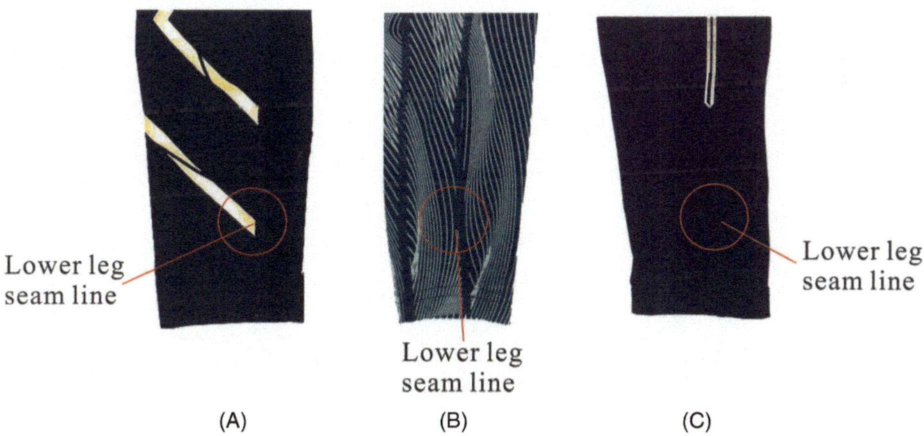

Figure 6.10 Seam locations at lower part of leg: (A) Sample II, (B) Sample III, and (C) Sample IV.

seams result in the reduction of garment-skin pressure in Samples II, III, and IV (Fig. 6.11). The flatlock seam used for the dual fabric layers in the interlace leggings adversely affects the contact conditions, thus leading to lower compression pressure as compared to Sample I which has no seam.

The influence of body movements on the compression pressure was also investigated. Compared to static standing, standing on a single leg (with the left leg off the ground) led to increased contraction of the rectus femoris muscle and ultimately, a

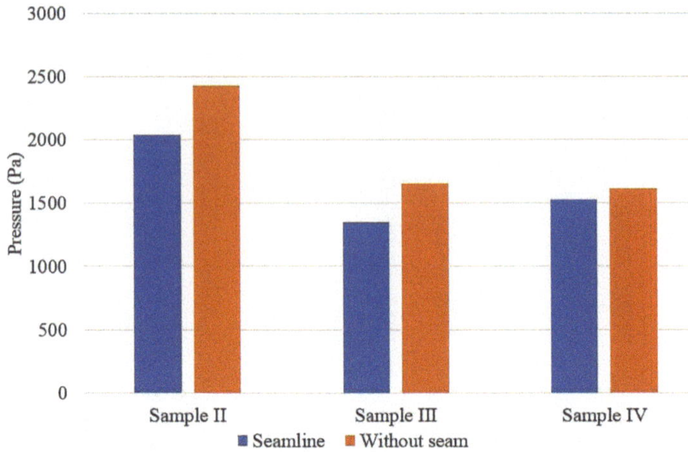

Figure 6.11 Pressure at lower part of leg with and without seamline – Samples II, III, and IV.

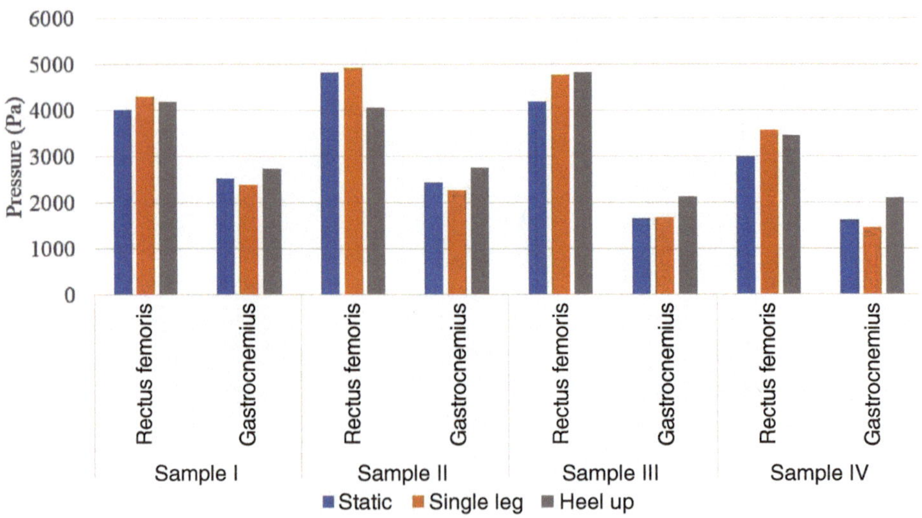

Figure 6.12 Pressure at rectus femoris and calf muscles measured for three different movements.

2%–20% increase in the compression pressure; see Fig. 6.12. The relaxation of the gastrocnemius muscle in the single leg stance also reduced the compression pressure. Amongst the four different legging samples studied, Sample II has the least amount of change in pressure in the rectus femoris muscle in the single leg stance. The engineered flocking pattern placed on the compression tights tends to reduce muscle movement and stabilize the targeted muscles. When the heels are turned upwards or the third body movement, the gastrocnemius muscle contracts to support the body weight, which results in a 9%–30% increase in the compression pressure.

It was observed that even though the subjects all have the same garment size, there are variations in body weight, height, waist and leg circumferences, as well as body shape geometry. They also have posture differences which may have affected the pressure measurements, and would have resulted in large variations in the compression pressure so that the overall mean tights-skin pressure ranges from 1000 Pa to 2663 Pa amongst the 10 subjects in this study.

6.4 Problems in designing compression activewear

Even though there has been an increasingly larger demand for compression activewear in the global sports apparel market, scientific evidence related to the design of compression activewear and the corresponding amount of pressure exerted has been largely absent. The pressure applied to the body is not only affected by the mechanical stress–strain properties of the textile materials, but also closely associated with the large variations in body curvature and pressure at different anatomic sites, deterioration of fabric elasticity, as well as the corresponding pressure and pattern engineering techniques.

6.4.1 Variations in body shapes

Gaied, Drapier, and Lun (2006) showed that the interface pressure produced by compression garments is affected by the geometry of the anatomical zones and the radius of the curvature of the anatomical body parts. Specific information on the geometric characteristics of the human body is, therefore, a critically important prerequisite to precisely determine the garment–skin interface pressure and the contact conditions of compression garments in relation to different body parts and joints. Three-dimensional body scanning technology only provides a better understanding of the complex and precise 3D characteristics of the human body, including the shape, size, mass, volume, etc. The concave areas of the body, for example, the waist and the web space between fingers, do not come into contact with compression garments, so it is difficult to apply adequate pressure in those places. As for compression swimsuit and tights, the bony pelvis and its rigidity bear most of the tension forces induced by a garment, which results in higher interface pressure and wear discomfort. As indicated by Tanaka, Midorikawa, and Tokura (2006), excessive high pressure induced on body parts may also cause numbness, or even breathing difficulties and other serious health issues. An anatomically engineered design of compression activewear, therefore, remains elusive.

6.4.2 Changes in body movements

Apart from the shape of the different anatomical zones, the pressure delivered by a compression garment is greatly affected by body posture and body movement during exercise. Body movement and posture are strongly related to skin deformations, particularly lower limb movements, thus resulting in garment distortion. The literature on the skin strain distribution of the lower extremities have indicated that the thigh has significantly larger stretch ratios than the calf and hips in running tasks (Wang & Wang, 2015). Since body shape can only be measured or scanned during static standing by using 3D body scanning technology, inherent ambiguity of the body dimensions is inevitable, and the geometry of the 3D body changes during movement and physical activity.

Taking a pressure therapy glove as an example, the skin–glove interface pressure can increase from 3.4 kPa in the relaxed state to 9.8 kPa with a clenched fist, which far exceeds the effective pressure for treating hypertrophic scars of 3.3 kPa (Yu, Yick, Ng, & Yip, 2015a; Yu, Yick, Ng, & Yip, 2015b). The high interface pressure can have detrimental effects on the skin oxygen pressure, which result in ischemia, bruises, and pressure ulcers. Comprehensive and robust 3D data from dynamic scanning are, therefore, important to enhance the design and fit of compression garments for optimum pressure control.

6.4.3 Loss of fabric elasticity

The elasticity of fabric is reduced after repeated wear. Fabric tension is gradually lost over time and with use, thus leading to pressure degradation and in terms of compression garments, reduced effectiveness. Yet the dynamic pressure distribution and pressure degradation have been generally neglected or over-simplified in most of the related studies (Ishimaru et al., 2011). For instance, Yu, Yick, Ng, and Yip (2013) used a fabric extender to simulate the repeated use of gloves in order to evaluate reductions in the elasticity of the tested gloves with three reduction factors of 10%, 15%, and 20%. The fabric strain, structure, and initial tension were identified as the key factors that affect the loss of fabric tension. Wang, Cui, Zhang, Feng, Shen, and Xiong (2011) investigated the loss of dynamic pressure in cotton/spandex fabrics with plain and rib knitted structures. They showed that the spandex feed rate and fabric structure primarily affect fabric pressure with repeated extension and recovery. It is anticipated that the choice of fabric, its structure, anisotropic behavior, knitting parameters, as well as the reduction factor in pattern engineering contribute to the loss of pressure during wearing.

6.4.4 Evaluation of garment pressure

Sustained pressure that exceeds 0.18 kPa in compression garments can result in complications to the body. For instance, studies have found that there can be damage to the mandible and thoracic cage area (Gallagher et al., 1992; Mann, Yeong, Moore, Colescott, & Engrav, 1997). In this respect, proper measurements and evaluation of the amount and distribution of the applied skin pressure are paramount in the design of compression garments.

Direct measurements of garment pressure are the not only conventional means of determining the interface pressure between the garment and skin, but also time consuming and intrusive. Also, when measuring the garment pressure, the testing devices

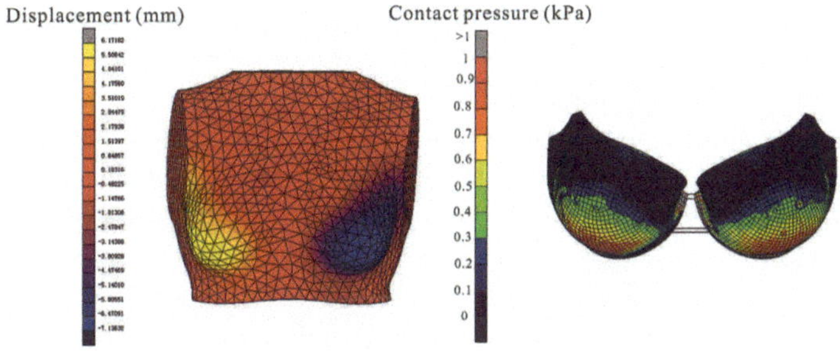

Figure 6.13 Dynamic simulation of breast displacement and contact pressure between breasts and bra.

might be affected by the different conditions of the body. As a result, some of the previous works have predicted the pressure exerted onto limbs through simulation by using pressure sleeves in a tubular form (Hui & Ng, 2003; Leung, Yuen, Ng, & Shi, 2010; Liu et al., 2005, 2006; Ng & Hui, 1999). Macintyre and Ferguson (2013) developed a tool to determine the amount of pressure induced onto limbs (which is limited to use on cylindrical shaped body parts) with Laplace's equation. The finite element (FE) approach which is widely used in biomechanics has also been adopted to simulate the biomechanical interactions between the human body and various loading conditions without the need for repeated wear trials, or experimental and product testing. The mechanical contact between various textile assemblies such as compression stockings, pants, and bras, and skin has been modeled to determine the stress–strain performance of the textile assemblies and analyze their pressure distribution on the body parts (Eder et al., 2013; Lee, Hong, & Lee, 2017; Lin, Choi, Luximon, Yao, Hu, & Li, 2011; Ng, Yu, & Li, 2009; Lu, Qiu, Wang, & Dai, 2016; Wang & Wang, 2015). However, most FE analyses assume that a rigid dummy or cylinder adequately represents the body, which neglects the complexities of the human body. Therefore, their estimated stress distributions are somewhat inaccurate. More recently, a 2D FE model of the cross section of the lower leg was used to identify the properties of skin, soft tissues and bones and explain for the relationship between pressure and displacement on the lower leg in relation to the tightness of socks (Dan, Fan, Shin, & Zhang, 2016). The FE method has been used to determine the dynamic displacement of the breasts caused by wearing a bra, as well as contact pressure between the body and girdle and breasts and bra based on the nonlinear properties of the soft tissues of breasts (Sun et al., 2019); see Fig. 6.13.

6.4.5 Subjective perception of garment discomfort

The pressure threshold that causes discomfort is about 5.88–9.80 kPa (Denton, 1972). However, this value varies with the individual person and the part of the body concerned, which could be similar to the average capillary blood pressure of 4.30 kPa near the skin surface. The pressure comfort zone in normal conditions is

1.96–3.92 kPa, but it also depends on the individual condition of the treated body part and body position.

Compression garments that induce high levels of pressure tend to affect psychophysical response, muscle activity, physical sensitivity, and perceived comfort of the wearers. As indicated by Yu et al. (2015a, 2015b), the use of pressure gloves not only results in a reduction in the maximum gripping force, but also adversely affects the perceived comfort and ease of hand motion. The results showed that even though both the fabric type and reduction factors of pressure gloves have no significant effects on the tactile sensitivity of the fingertips, the active range of motion and dexterity of the fingers in carrying out daily tasks and perceived comfort are considerably affected. The use of a high reduction factor of 20% in glove patterns can negatively impact both hand functions and perceived comfort. Strong associations were found between the comfort performance and fabric properties, including surface roughness, bending rigidity, thermal conductivity, and moisture retention.

The use of compression garments is likely to increase the skin temperature, especially during intense physical exercise that incurs heavy sweating. The capability of compression textile materials to regulate temperature and humidity, therefore, has a decisive influence on the garment-skin microclimate and thermal comfort. The design of compression activewear such as the construction of the garment, fit, and material (viz., layers, thickness, and porosity) that affect the ability to maintain a layer of still air in the garment has major effects on the garment insulation properties and ventilation. Depending on the end-use and wear conditions, some compression activewear dresses are designed to keep the body warm in cold environments, whereas other types of activewear are designed to facilitate the evaporation of heat caused by exercise or sports activities. Knowledge on the type of compression material and design features, and their relationship with the thermal and moisture distributions of the body and subjective perceptions of compression activewear could provide useful guidelines that advance the design of compression activewear with better wear comfort.

6.5 Conclusion

A compression activewear can apply mechanical pressure to the limbs, or to the upper and lower parts of the body, or the entire body, to compress and support the underlying tissues and muscles which prevent potential injury, especially during high intensity activity. The major advantages of compression activewear include improved circulation, attenuation of muscular oscillations, decrease in muscle damage, and reduced time for recovery of muscle injuries. However, there has been very little work to date in the literature that provides accurate and objective techniques to evaluate and quantify the true effectiveness of compression activewear. Scientific research and reliable methods that examine the body shape, movement, fabric properties, and garment pressure could facilitate improvements to compression activewear, and provide a better understanding of the material selection and means of optimizing the development and design of such garments.

References

[2XU] (2019). Women. Retrieved from https://www.2xu.com/us/women-2/.

Adidas (2019). Alphaskin sport tights. Retrieved from https://www.adidas.com/us/alphaskin-sport-tights/CF7339.html.

Arena (2019). Women's carbon compression long sleeve. Retrieved from https://www.arenawaterinstinct.com/en_global/women-s-carbon-compression-long-sleeve.html.

CW-X (2019). CW-X patented support technology, EXO-WEB technology: bottoms. Retrieved from https://cw-x.com/technology/exo-web-technology-bottoms/.

Dan, R., Fan, X., Shin, Z., & Zhang, M. (2016). Finite element simulation of pressure, displacement, and area shrinkage mass of lower leg with time for the top part of men's sock. *Journal of the Textile Institute*, *107*(1), 72–80.

Denton, M. (1972). Fit, stretch and comfort. *Textiles*, *3*, 12–17.

Eder, M., Raithm, S., Jalali, J., Volf, A., Settles, M., Machenss, H. G., & Kovacs, L. (2013). Comparison of different material models to simulate 3D breast deformations using finite element analysis. *Annals of Biomedical Engineering*, *42*(4), 843–857.

Gaied, I., Drapier, S., & Lun, B. (2006). Experimental assessment & analytical 2D predictions of stocking pressures induced on a model leg by compressive stocking. *Journal of Biomechanics*, *16*, 3017–3025.

Gallagher, J., Kaplan, S., Maguire, G. H., Leman, C. J., Johnson, P., & Elbaum, L. (1992). Compliance and durability in pressure garments. *Journal of Burn Care & Rehabilitation*, *13*(2), 239–243.

Halseth, T., McChesney, J. W., DeBeliso, M., Vaughn, R., & Lien, J. (2004). The effects of kinesio taping on proprioception at the ankle. *Journal of Sports Science and Medicine*, *3*(1), 1–7.

Huang, C. Y., Hsieh, T. H., Lu, S. C., & Su, F. C. (2011). Effect of the Kinesio tape to muscle activity and vertical jump performance in healthy inactive people. *Biomedical Engineering Online*, *10*(1), 1.

Hui, C. L., & Ng, S. F. (2003). Theoretical analysis of tension and pressure decay of a tubular elastic fabric. *Textile Research Journal*, *73*(3), 268–272.

Ishimaru, S., Isogai, Y., Matsui, M., Furuichi, K., Nonomura, C., & Yokoyama, A. (2011). Prediction method for clothing pressure distribution by numerical approach: Attention to deformation by extension of knitted fabric. *Textile Research Journal*, *81*(18), 1851–1870.

Lee, H., Hong, K., & Lee, Y. (2017). Compression pants with differential pressurization: Kinetic and kinematical effects on stability. *Textile Research Journal*, *87*(3), 1554–1564.

Lemos, T. V., Pereira, K. C., Protássio, C. C., Lucas, L. B., & Matheus, J. P. C. (2015). The effect of kinesio taping on handgrip strength. *Journal of Physical Therapy Science*, *27*(3), 567–570.

Leung, W. Y., Yuen, D. W., Ng, S. P., & Shi, S. Q. (2010). Pressure prediction model for compression garment design. *Journal of Burn Care & Research*, *31*(5), 716–725.

Lin, Y., Choi, K. F., Luximon, A., Yao, L., Hu, J. Y., & Li, Y. (2011). Finite element modeling of male leg and sportswear: Contact pressure and clothing deformation. *Textile Research Journal*, *81*(14), 1470–1476.

Liu, H., Chen, D., Wei, Q., & Pan, R. (2011). A study of relationship between clothing pressure and garment bust strain, Young's modulus of fabric, based on a FE model. *Textile Research Journal*, *81*(13), 1307–1319.

Liu, R., Kwok, Y. L., Li, Y., Lao, T. T., & Zhang, X. (2005). Effects of material properties and fabric structure characteristics of graduated compression stockings on the skin pressure distribution. *Fibers and Polymers*, *6*(4), 322–331.

Liu, R., Kwok, Y. L., Li, Y., Lao, T. T., Zhang, X., & Dai, X. Q. (2006). A three-dimensional biomechanical model for numerical simulation of dynamic pressure functional performances of graduated compression stocking (GCS). *Fibers and Polymers*, *7*(4), 389–397.

Liu, R., Kwok, Y. L., Li, Y., Lao, T. T., & Zhang, X. (2007). Quantitative assessment of relationship between pressure performances and material mechanical properties of medical graduated compression stockings. *Journal of Applied Polymer Science*, *104*, 601–610.

Liu, R., Guo, X., Lao, T. T., & Little, T. (2017). A critical review on compression textiles for compression therapy: Textile-based compression interventions for chronic venous insufficiency. *Textile Research Journal*, *87*(9), 1121–1141.

Liu, Y., Hu, H., Zhao, L., & Long, H. (2012). Compression behavior of warp-knitted spacer fabrics for cushioning applications. *Textile Research Journal*, *82*(1), 11–20.

Lu, M., Qiu, J., Wang, G., & Dai, X. (2016). Mechanical analysis of breast-bra interaction for sports bra design. *Materials Today Communications*, *6*(2016), 28–36.

Macintyre, L., & Baird, M. (2006). Pressure garments for use in the treatment of hypertrophic scars – A review of the problems associated with their use. *Burns*, *32*, 10–15.

Macintyre, L., & Ferguson, R. (2013). Pressure garment design tool to monitor exerted pressures. *Burns*, *6*, 1073–1082.

MacRae, B. A., Cotter, J. D., & Laing, R. M. (2011). Compression garments and exercise: Garment considerations, physiology and performance. *Sports Medicine*, *41*, 815–843.

Mann, R., Yeong, E. K., Moore, M., Colescott, D., & Engrav, L. H. (1997). Do custom-fitted pressure garments provide adequate pressure? *Journal of Burn Care & Rehabilitation*, *18*(3), 247–249.

Manshahia, M., & Das, A. (Dec 2014). High active sportswear – A critical review. *Indian Journal of Fibre & Textile Research*, *39*, 441–449.

Mizuno, S., Arai, M., Todoko, F., Yamada, E., & Goto, K. (2017). Wearing lower-body compression garment with medium pressure impaired exercise-induced performance decrement during prolonged running. *PLOS One*, *12*(5), e0178620.

Morooka, H., Nakahashi, M., Morroka, H., & Kitamura, K. (2001). Effects of clothing pressure exerted on a trunk on heart rate, blood pressure, skin blood flow and respiratory function. *Journal of the Textile Machinery Society of Japan*, *54*, 57–62.

Ng, F. S. F., & Hui, P. C. L. (1999). Effect of hem edges on the interface pressure of pressure garments. *International Journal of Clothing Science & Technology*, *11*(5), 251–261.

Ng, S. P., Yu, W., & Li, Y. (2009). Photogrammetric prediction of girdle pressure. *Measurement Science and Technology*, *20*(1), 015804.

NIKE (2019). NIKE Pro Hypercool compression. Retrieved from https://m.nike.com.hk/product/636157-010/detail.htm.

Rajan, T. P., Souza, L. D., Ramakrishnan, G., & Zakriya, G. M. (2016). Comfort properties of functional warp-knitted polyester spacer fabrics for shoe insole applications. *Journal of Industrial Textiles*, *45*(6), 1239–1251.

Shishoo, R. (2015). Knitting technology for seamless sportswear. In *Textiles for sportswear* (pp. 95–117), Cambridge: Woodhead Publishing.

SPEEDO (2019). Fastskin LZR intent openback. Retrieved from https://www.speedo.com/international/en/fastskin-lzr-intent-openback/811974D168.html.

Sperlich, B., Born, D. P., Kaskinoro, K., Kalliokoski, K. K., & Laaksonen, M. S. (2013). Squeezing the muscle: Compression clothing and muscle metabolism during recovery from high intensity exercise. *PLOS One*, *8*(4), e60923.

Statista (2019). Projected development of the global sports apparel market revenue from 2018 to 2024. Retrieved from https://www.statista.com/statistics/412667/global-sports-apparel-market-projected-development/.

Sun, Y., Yick, K. L., Yu, W., Chen, L. H., Lau, N., Jiao, W. Z., & Zhang, S. C. (2019.). 3D bra and human interactive modelling using finite element method for bra design. *Computer-Aided Design.*, *114*, 13–27.

Tanaka, S., Midorikawa, T., & Tokura, H. (2006). Effects of pressure exerted on the skin by elastic cord on the core temperature, body weight loss and salivary secretion rate at 35 °C. *European Journal of Applied Physiology*, *96*, 471–476.

Vered, E., Oved, L., Zilberg, D., & Kalichman, L. (2016). Influence of kinesio tape application direction on peak force of biceps brachii muscle: A repeated measurement study. *Journal of Bodywork and Movement Therapies*, *20*(1), 203–207.

Wang, Y., Cui, Y., Zhang, P., Feng, X., Shen, J., & Xiong, Q. (2011). A smart mannequin system for the pressure performance evaluation of compression garments. *Textile Research Journal*, *81*(11), 1113–1123.

Wang, Y., & Wang, J. (2015). Design data for running tight: Skin strain distribution on lower extremity based on decomposition of movement. *Journal of the Textile Institute*, *106*, 469–479.

Xiong, Y., & Tao, X. M. (2018). Compression garments for medical therapy and sports. *Polymers*, *10*, 663.

Yasukawa, A., Patel, P., & Sisung, C. (2006). Pilot study: Investigating the effects of kinesio taping® in an acute pediatric rehabilitation setting. *American Journal of Occupational Therapy*, *60*(1), 104–110.

Yu, A., Yick, K. L., Ng, Z., & Yip, J. (2015a). The effect of pressure and fabrication of pressure therapy gloves on hand sensitivity and dexterity. *Journal of Burn Care & Research*, *36*(3), 162–175.

Yu, A., Yick, K. L., Ng, Z., & Yip, J. (2015b). The effect of pressure glove tightness on forearm muscle activity and psychophysical responses. *Human Factors*, *57*(6), 988–1001.

Yu, A., Yick, K. L., Ng, S. P., & Yip, J. (2013). Prediction of fabric tension and pressure decay for the development of pressure therapy gloves. *Textile Research Journal*, *83*(3), 269–287.

3D printing technologies: current applications, future trends, and challenges

7

Lushan Sun
Institute of Textiles and Clothing, The Hong Kong Polytechnic University, Hung Hom, Hong Kong

7.1 Introduction

Additive manufacturing has been in use for over three decades and is recognized in many industries for its fabrication and design efficiency which has resulted in a substantially reduced lead time in production as well as versatility in applications. In its early days, additive manufacturing was commonly known as rapid prototyping. Today, it is greatly well known by its umbrella terms, including three-dimensional printing (3DP) and direct digital manufacturing (DDM) (Sun & Zhao, 2017). The increasing popularity of 3DP technologies have resulted in their exponential advancement so that they now provide products with the quality of those that are mass manufactured and mass customized. The printing methods have since advanced to create higher quality output (e.g., fused deposition modeling (FDM) for filament printing, selective laser sintering (SLS) for powder printing, and PolyJet for multi-material printing with resin).

In the early days of 3D printed wearables, the fashion industry explored its applications for accessory products and conceptual art designs and customization. With the maturity of 3D printing technologies, increasingly more designers are now pushing the boundaries by using 3D printing to create materials and 3D computer aided design (CAD) modeling methods to design more wearable or ready-to-wear apparel. Designers have since found that 3D printing technologies can be modified for apparel, thus allowing process and workflow disruption, or even hacking of the technologies, so that it is now important to rethink traditional skill sets and evolve them. Although 3D printing applications are still in their infancy, this chapter focuses on a few cases around activewear apparel and footwear with the use of 3D printing as they have been unique in their development approaches, material uses, and applied technologies. The intention is to shed light on the challenges and the potential future integration of 3D printing technologies in the activewear industry.

7.2 Integrating 3D printing with 3D CAD modeling

Unlike the early pioneers of 3D printed fashion who focused on dramatic art form and structural complexity and novelty, designers today who hail from diverse fields are truly seeking ways to disrupt the conventional views around apparel design and

Latest Material and Technological Developments for Activewear. http://dx.doi.org/10.1016/B978-0-12-819492-8.00007-7

practices in order to facilitate the true potential of 3DP in apparel. The integration of 3DP in conventional ready to wear apparel often focuses on manipulating 3D printed structures for increased material flexibility during wear, and selecting the appropriate type of material and 3DP method for aesthetic and functional outcomes. However, it is more important to achieve wear comfort in designing activewear apparel and footwear for the intended activities with the use of such technologies.

7.2.1 Fully 3D printed activewear with direct 3D CAD modeling

In 2014, Danit Peleg debuted the world's first fully 3D printed apparel collection by using FDM to print filaments. Her work has since evolved into a 3DP business for customized apparel. Her latest collection, The Birth of Venus, showcases another stunning series of 3D printed outfits (DANIT PELEG, 2018). Among the featured outfits is one of the most stunning and most versatile ready to wear jackets that can be mixed and matched with conventional apparel (Fig. 7.1).

Peleg's background is in traditional fashion design. In her design process with the use of 3DP, she often starts by designing the "textiles" that complement the intended functions of the garment and their visual impact. The front zip-up jacket in The Birth of Venus collection starts with the basic unit of a "textile" pattern or 3D printed structure. The zigzag pattern is sourced online and then manipulated by using 3D CAD modeling software, which is a common technique called direct modeling that enables various segments of the unit to be replicated and joined together digitally to fabricate a meter of "textile". This zigzag pattern is joined in a way that allows four-way stretch (Fig. 7.1A), thus allowing flexibility in the "textile" and even creating a beautiful drape. The digital modeling process also enables the designer to further control and manipulate the performance and behavior of a pattern by changing not only its scale and density of weave repetition, but also the thickness and position of the unit pattern. Moreover, having a good understanding of the potential 3D printed materials and printing methods used is critical in controlling the design outcome.

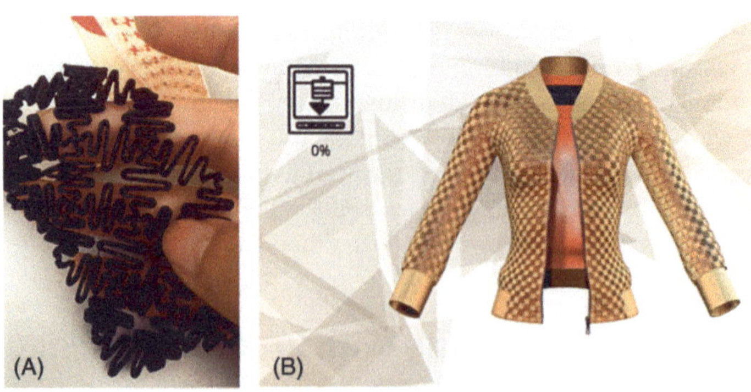

Figure 7.1 (A) Virtual simulation of the ready to wear jacket by Danit Peleg from The Birth of Venus collection (2018) for consumer online customization and (B) sample flexible TPU textile printed using opensource 3D CAD model and FDM 3D printer.

Currently, Peleg has been focusing on using a commercial grade FDM 3D printer that produces Filaflex filaments, which is a type of thermoplastic polyurethane (TPU) with a rubbery hand (Fig. 7.1A). Her garment designs can be summed up as four main steps, including "textile" designing, 3DP of the "textile" sections, assembly, and post-finishing. Each step needs to be considered in the digital designing and iterating processes throughout the development of the garment. Today, Peleg's customization business offers consumers an online platform to visualize and order their garment based on preferred color combinations, personalized text, and preset size via a mobile-based virtual fitting app called Neitelo (Fig. 7.1B) (DANIT PELEG, 2018).

7.2.2 Fully 3D printed activewear with computational 3D modeling

The technology of 3D CAD modeling can be enhanced with algorithm-based computational design features and become very useful in 3D fully printed activewear designs. The N12 bikini, developed by the co-founders of Continuum Fashion—Mary Huang, a media artist whose expertise is in interaction design, and Jeena Fizel, an architect who specializes in programming interactive environments, is currently the world's first fully 3D printed ready-to-wear swimwear (Fig. 7.2). SLS is used to construct the bikini, which is a printing technology that applies a laser to sinter powdered material (Continuum Fashion, n.d.). Therefore, the bikini is fabricated by 3DP laser fused nylon powder, which is also known as N12. The resultant printed material has a highly even surface, durable, and light in weight, and can have a wall thickness as small as 0.7 mm (Continuum Fashion, n.d.). N12 is also innately waterproof and even more comfortable to wear when it becomes wet (N12, n.d.).

The nylon powder for the bikini is printed in shape of, what the designers call, circular packing system. This system includes plates that vary in size are connected with strings so that the bikini is not only strong but also flexible (Fig. 7.2C). In the digital design process, a customized algorithm code was written for the pattern to help produce printed materials with a structure that can adapt to a preset surface or

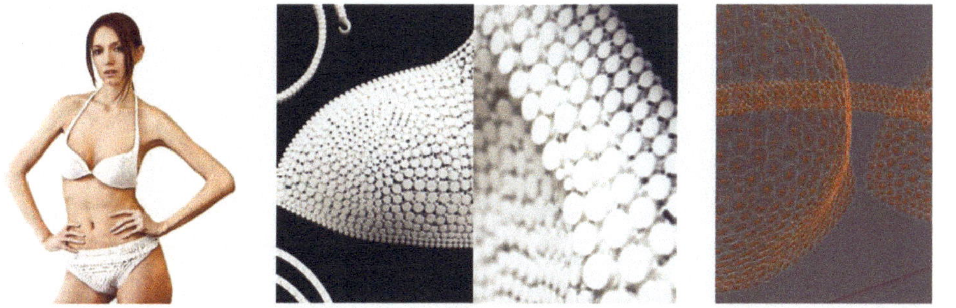

(A) (B) (C)

Figure 7.2 (A) N12 Bikini by Continuum Fashion, (B) structure details of bikini and (C) details of packed circular plates in 3D CAD program.

the body. That is, the circular plates can be laid out to accommodate the curves of the body. Therefore, the aesthetics of this garment are entirely derived from its structural design (N12, n.d.). This computational design technique enables the fabrication of complex and intricate 3D printed structures which would otherwise be impossible by hand or using other conventional methods. The technique also creates a system that can respond continuously to the body or the shape that is being covered (N12, n.d.). Additionally, the bikini top comes with modular halter straps, and there are hooks at the center of the front and back of the bikini top for closure. For the bottom part of the bikini, a belt is digitally designed at the lower waist area and connected to the crotch sections. Currently, Continuum Fashion offers the N12 bikini online with color options of black or white.

7.2.3 3D partially printed activewear with direct 3D CAD modeling

Although 3DP can be used to fully fabricate a garment, in some cases, it may be more practical to partially integrate 3DP so that both traditional textiles and 3D printed materials can be used together. Fig. 7.3 shows two examples of activewear from a design research case study (Cui & Sun, 2018; Cui, 2019), developed by academics with industrial and apparel design backgrounds. The goal of these design case studies was to

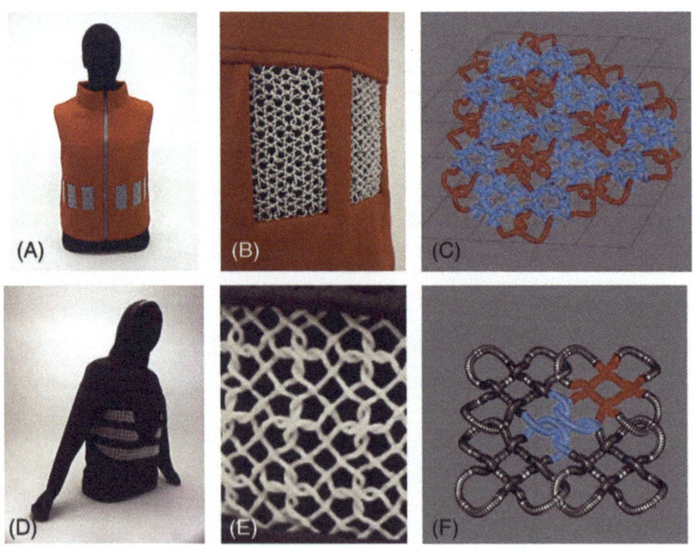

Figure 7.3 (A) Hybrid Comfort—beach vest for men in neoprene knit and filament 3D printed TPU vents (Cui & Sun, 2018), (B) flexible 3D printed structure of vest, (C) three-way and six-way structure using 3D CAD program (Cui & Sun, 2018), and (D) Infinite Vitality—unisex hooded sweatshirt with fleece knit with filament 3D printed TPU vents, (E) flexible 3D printed structure of sweatshirt, and (F) four-way loop structure using 3D CAD program (Cui, 2019).

examine the knowledge required to design and integrate filament 3D printed structures that are TPU based with traditional knits. In both cases, the activewear function was used as a focal point for integrating the 3D printed parts.

Hybrid Comfort (Fig. 7.3A) is a beach vest for men fabricated in a neoprene knit, which has filament 3D printed TPU vents in the front abdomen and upper back areas. The resulting 3D printed structure has also incorporated direct 3D CAD modeling, in which a basic unit is used throughout the 3D model, and then repeated to form a larger motif (Fig. 7.3C). By looping segments, the 3D model here shows a triangular shaped or three-way unit in red and a hexagon shaped or six-way unit in blue. They are joined repeatedly to form a rectangular pattern that is approximately 3 mm in total thickness. Although there is thickness, the rectangular section provides an appropriate balance between flexibility and structured durability for the abdominal part of the beach vest. It is also appropriate and feasible for the thickness of the neoprene chosen and sewing techniques applied. To ensure a smooth seam finish, the 3D model includes a frame around the rectangular pattern that measures 0.5 cm in width and 0.15 mm in thickness for the seam allowance in sewing. Straight stitching and topstitching are carried out on a sewing machine for the vest. One or two layers of neoprene knit are stitched with one flat layer of 3D printed TPU material.

Infinite Vitality is a unisex hooded sweatshirt made of fleece knit and FDM-printed TPU vents (Fig. 7.3E). This design study was a pilot study to the Hybrid Comfort vest which was intended to explore and evaluate the performance and feasibility of integrating 3D printed structures for knit-based activewear. In considering the donning and doffing of a hooded sweatshirt, the designers paid particular attention to the level of comfort with the use of 3DP. The 3D printed structures are a hybrid twisted woven pattern, repeated uniformly for all of the venting strips (Fig. 7.3D–F). Here, the 3D printed structure was developed by exploring the variations in conventional woven textile structures. The basic unit is uniformly repeated in a block pattern and a sample was 3D printed for performance and comfort evaluation. In the final prototype, the 3D printed insets were engineered to specific dimensions based on the flat pattern of the garment. They visually resemble diamond tiles, developed from a four-way unit repeated pattern (Fig. 7.3F). Also, compared to the structure in the Hybrid Comfort vest, they are designed to have less bulk for structural flexibility and resilience.

Furthermore, the level of difficulty in partially integrating complex 3D printed parts increases with curved garment seams. The bell-shaped curve of the seam in the front of the bodice intersects with four pieces of 3D printed venting strips on both sides of the sweatshirt. In creating fluidity, the garment design does not have any side seams but accommodates the venting insets with curved seams under the bottom strip that descend toward the backside of the sweatshirt. Because of the material and seam complexities, the 3D printed insets are constructed with knitted fleece parts with zig-zag top-stitching to provide a polished look. Unlike the relatively static 3D printed vents in the Hybrid Comfort vest, no frames are designed around the 3D printed insets on the sweatshirt to allow flexibility during wear. Lastly, it is important to point out that the 3D printed insets in both garments require post-finishing processes to clean or remove minor supporting material from the typical 3DP of filaments based on the FDM.

7.2.4 3D printing integrated footwear with computational 3D modeling

The world's leading footwear makers also use 3DP technologies as they find that the technologies allow the creation of unique and complex footwear structures. In 2016, Under Armour (UA), as the first global brand to market 3D printed footwear, released its second generation of the ArchiTech footwear series for men, the Futurist line. Developed in its lab called the Lighthouse, the Futurist line continues to push the boundaries of 3DP technology in the development of footwear soles. The Futurist line also features a compression lacing system for the shoe upper, which uses UA SpeedForm technology, to help mold the foot and eliminate distractions during training. Inspired by traditional garment patterns, the designers engineered external sleeve wraps from the midfoot to ankle for extended support (Fig. 7.4).

The mint green midsole of the Futurist footwear is 3D printed by SLS with TPU powder (Fig. 7.4). It is partially integrated into the shoe to create a stable platform that provides adequate heel support, cushioning, and durability for strength training and lifting. In accordance with advanced computational 3D modeling techniques, the designers and engineers at UA experimented with different complex structure formations to ensure the most effective performance of the midsole. Another critical factor in the testing is the lacquer coating on the 3D printed midsole. This proprietary technology further enhances the durability and resilience of the structure to optimize the shoe performance. Overall, the Futurist footwear is both a highly functional and aesthetically fashionable item for individuals who are interested in sports and recreation.

7.2.5 3D fully printed footwear with computational 3D modeling

Three-dimensional printing allows for the creation of incredibly complex structures which are customizable for each wearer (Jamie, 2017). An industrial designer by the name of Earl Stewart designed the XYZ shoe which was produced by using 3DP and

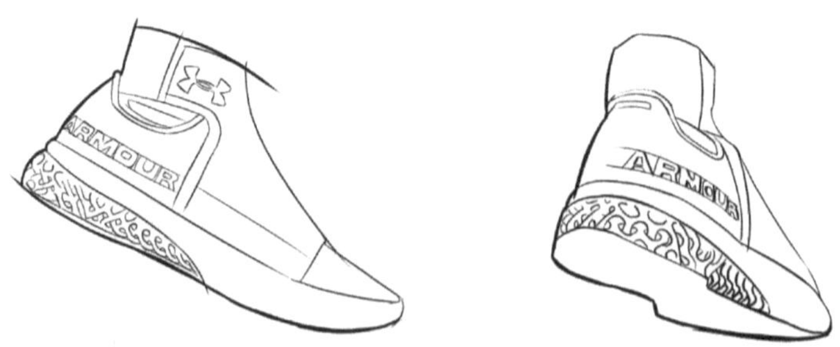

Figure 7.4 Sketches of Under Armour ArchiTech Futurist training shoes for men, featuring 3D printed mid-sole using lattice structures.

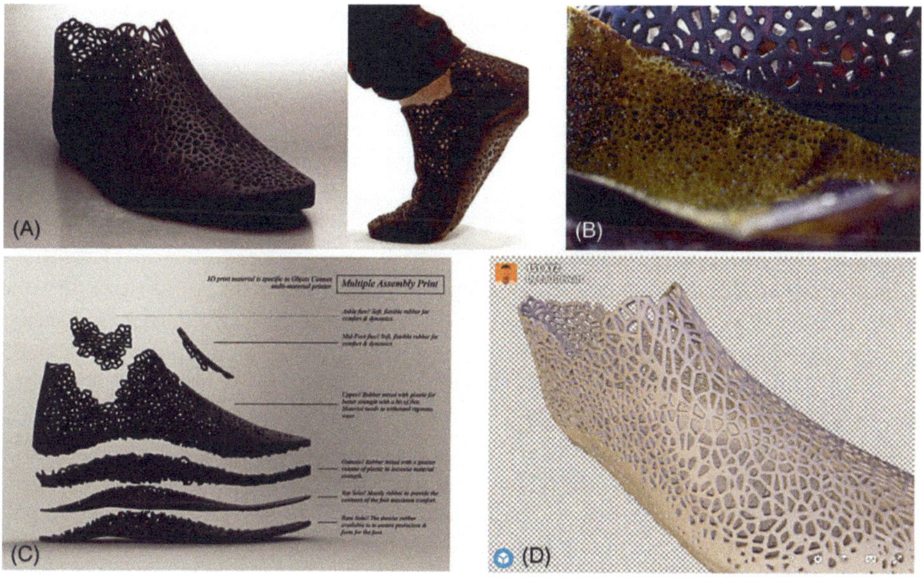

Figure 7.5 (A) 3D fully printed footwear—XYZ shoe, (B) multi-material details, (C) multi-material jetted layer breakdown, and (D) 3D CAD model (XYZ Shoe, 2013).

explores design pluralism by considering biological and cultural values (Fig. 7.5). The design goal was to build a fully 3D printed shoe that caters to biomechanical needs. The XYZ shoe is fabricated by using the multi-material jetting technology, which is a resin-based method that seamlessly forms both flexible and rigid parts in one printing effort (Fig. 7.5B and C).

This design was developed in collaboration with a podiatrist for a shoe that provides more comfort and alignment stability with a custom fit based on the use of 3D foot scanning (XYZ Shoe, 2013). Better fit and comfort are achieved by using a three-layer sole system (Fig. 7.5C). The outsole is fabricated by using rubber mixed with rigid plastics for material strength. The top sole is a mixed material that aims to contour the foot. The base sole is made with a dense rubber mixture to provide foot protection. The shoe upper is constructed with both rigid and flexible materials for wear durability. Segments of the rubber material are placed at the mid-foot and flexion of the ankle for comfort.

In the digital design process, the designer applied computational 3D modeling to generate a structure that represents the biological growth from natural cell division which depicts one's cultural origins (XYZ Shoe, 2013) (Fig. 7.5D). Three-dimensional scanning of the foot is used in the 3D CAD modeling process to provide the data for the cell-like structure of the shoe, which is parametrically programmed to provide an ergonomic fit. A visual programming tool is experimentally used here to develop the most biomechanically effective and efficient density of the structure (XYZ Shoe, 2013).

7.3 Technology and material capabilities in activewear prototyping

As 3DP technologies diversify and advance, more materials and methods are becoming feasible for use in ready-to-wear apparel and even functional activewear. Thus far, most designers find three types of printing technologies to be the most advantageous for apparel and footwear. First is the use of SLS, which is a fusing process (Fig. 7.6) that relies on a focused laser beam to sinter powdered material (Lipson & Kurman, 2013). The bottom of the building chamber, which may vary in size, moves down as the object forms through sintering. Typically, a roller is used to continuously add and smooth each layer of sinter. Generally, SLS is time consuming, complicated, and costly, compared to other options. However, SLS can produce high-quality effects and structures that are highly complex and accurate, which cannot be realized with traditional or subtractive methods.

Furthermore, the same powdered material in SLS serves as both the building and supporting material. Often, much of the used powder, up to 70%–80%, can be recycled for the next printing job, thus, thus facilitating a more sustainable process. One critical aspect is that SLS printers have been typically built as a piece of industrial machinery which are large and expensive to maintain. In recent years, desktop SLS printers are made available to support small and customized production processes, at a fraction of the cost. It is important to note that SLS prints objects in a chamber, which means that 3DP can be carried in a free-flow position in the chamber. In some advanced digital design (Nervous System Inc., 2014), a large 3D CAD model with articulating joints can be collapsed digitally into a smaller mass and fabricated with an SLS machine in one print job.

As for the powdered material, SLS printers commonly print with different types of nylon or polyamide powder. In some cases, designers also use TPU powder for different types of material properties. Compared to melted material, the laser sintering process binds the powder together to form a more porous surface, as opposed to a glossy finish. The final prints usually need to undergo post-finishing processes to clean the surface and ensure a market grade product (Armstrong, 2019). After removing large chunks of loose powder, the object usually needs to be blown with compressed air

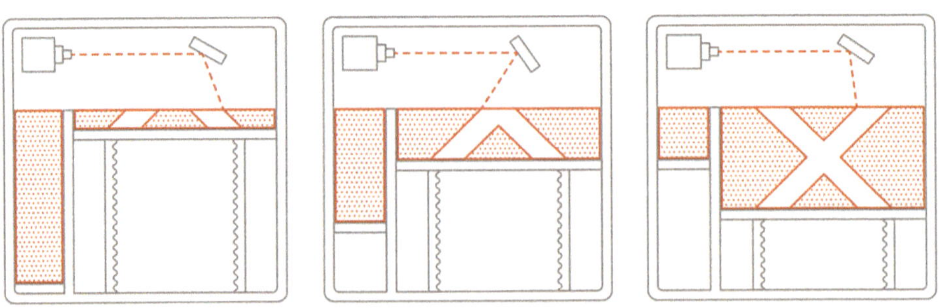

Figure 7.6 3D printing with SLS printer using powdered material.
Source: Varotsis (2019).

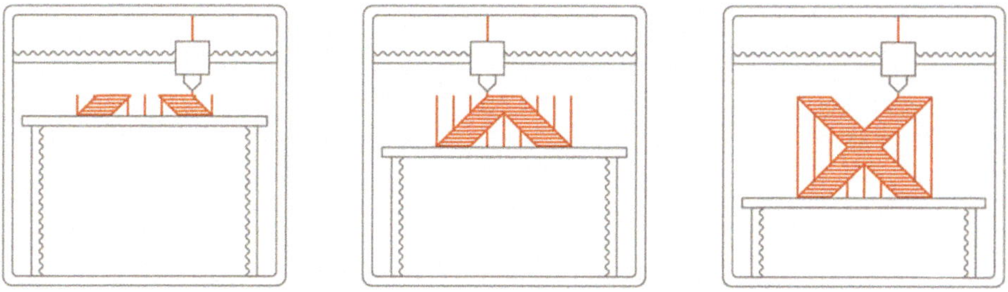

Figure 7.7 3DP with FDM method using filaments as building material.
Source: Varotsis (2019).

and blasted with plastic beads to remove any unsintered powder that is adhered to the surface. At this point, the surface is matte with a slightly grainy texture and best for applications that are conducive to synthetic dyes and paints.

The second most frequently used printing method for apparel design is FDM (Fig. 7.7). The fused deposition printer mainly deposits filament material through one or more nozzles and builds an object on a flat plate (Lipson & Kurman, 2013). The plate moves down, or the nozzle of the FDM printer moves up, as the printed object increases in height. The printing quality is generally based on the nozzle size and printing resolution of the printer, complexity, and position of the printed object, type and diameter of the filament, and different support structure applications. Often, complex or intricate objects, depending on their position, will require a thin layer of support material to hold the hollowed parts in place during the printing process. In some production grade FDM printers (e.g., Ultimaker), a dissolvable support material is used to enable free-flowing printing for complex structures.

As for feasible types of filaments, activewear designs frequently use different kinds of nylon and TPU. Both materials may come in softer and stiffer versions but provide more resilience and durability than other types of filaments, such as polylactic acid (PLA). The filaments for FDM are also available in a large number of color options. Today, apparel designers often consider FDM to be the most cost-effective, rapid, and versatile method of 3DP. This method gives designers greater capability of hacking conventional 3DP approaches for apparel development. Among them is the ability to easily modify the printing dimensions and adjusting the printing speed and temperatures for different filaments and designs. Most interestingly, some FDM systems can be modified to 3D print an object on top of another substrate or a piece of traditional textile by using unconventional or innovative materials such as flexible cellulose filaments (Technical Research Centre of Finland (VTT), 2017) and cell-based bio-materials (Tucker, 2015) for various functional applications of activewear.

Finally, in considering its flexibility and versatility, material jetting 3DP can be used as yet another option for fabricating activewear and related footwear. It is a technology that uses liquid resin in building material (Lipson & Kurman, 2013) (Fig. 7.8). Resin drops are selectively deposited and cured on a building plate. In the advanced material jetting 3D printers (e.g., PolyJet), different kinds of materials can be used in one print job with customized material characteristics. The surface of the resultant

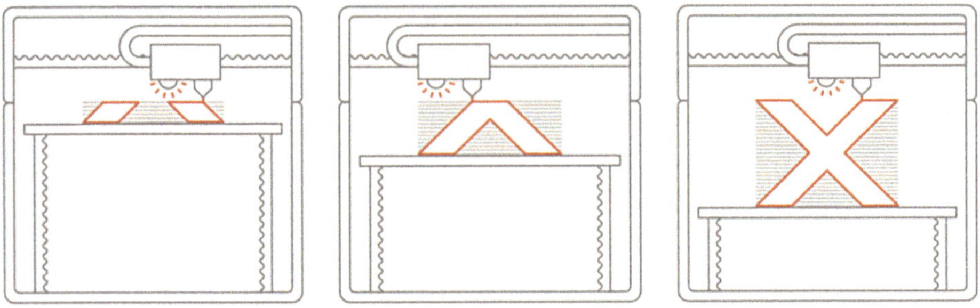

Figure 7.8 Material jetting 3D printing technology with liquid resin as building material and dissolvable or meltable support material.
Source: Varotsis (2019).

product is very smooth, and can be matte, glossy, or a mix of both textures. Although little post-finishing is required, material jetting technology requires the use of a separate support material in the printing process, one that is usually dissolvable or meltable. However, this increases the cost, preparation time, and structural integrity of some print jobs (Armstrong, 2019). The resin material may range from stiff PLA to the rubbery TPU, advanced thermoplastic composites with combined materials and varying percentages. Like FDM, there are many color options for material jetting. Another unique advantage of using this type of 3DP is that the method produces a more finished surface with higher resolution for market quality products. However, because of the complexity and the cost of the method, apparel designers have favored FDM over material jetting. Currently, it is more feasible for partially integrating 3DP to create activewear due to its limitations in building volume.

7.4 Advantages of 3D printing activewear and skills involved

Over the last 15 years of progress and growth, the apparel and textiles industry professionals and academics have come to recognize the potential and future possibilities of 3DP for various product categories (Delamore, 2004; Sisson & Thompson, 2012; Sun & Zhao, 2017). Activewear has been one of the most challenging apparel categories for integrating 3DP due to its demands for function and comfort. However, designers of fashion and those in other relevant fields, such as computer science, industrial and architectural design, and biological engineering, are now collaborating to carry out interdisciplinary designs and identify the key advantages and ways to overcome the critical challenges.

7.4.1 Design complexity and efficiency

One of the key advantages of using 3DP is to produce highly complex structures rather efficiently for two main reasons. First, digital 3D models can be easily scalable and modified for design variations. Designers can quickly change, reverse, improve, or save a design with a few clicks of the mouse when using 3D CAD tools. In some cases, 3D scanning technology can be used for reverse engineering a shape or surface from existing objects. For instance, 3D body scans can be used to provide the surface geometry information for a garment part or serve as a virtual dress form in 3D CAD modeling (Continuum Fashion, n.d.). In advanced processes, 3D simulation technology is useful in full or partial virtual garment fitting or simulate and articulate structure in the physical world (Nervous System Inc., 2014). The methods (e.g., SLS, FDM) discussed in this chapter will allow designers to have more freedom and more creative outlets in resolving design dilemmas and challenges.

7.4.2 Skills and knowledge for 3D printing activewear

The increased complexity of activewear also inevitably disrupts the conventional design flow. Many of the conventional ways focus on physical and tangible materials but the use of 3DP often starts with a virtual environment. The designer has to therefore take more into consideration the fundamental structure of the material in terms of

- how it influences the function of the final garment,
- how it visually impacts garment aesthetics and style lines,
- how it interacts with traditional textiles and other construction notions, such as zippers and elastic, and finally,
- how it provides comfort and maximizes performance for the wearer.

Ultimately, designers have to determine how they would adapt to new technology based on their existing knowledge and skill sets. Traditional apparel design training and thinking are derived from flat pattern making, draping, and sewing techniques, and builds upon them. Although digital technologies, such as digital textile design and knitting, have been applied in different activewear designs, they mostly focus on two-dimensional (2D) digital design techniques. With 3DP, the virtual environment requires *three-dimensional understanding and spatial visualization skills* (Sun & Zhao, 2017), hence converting 2D designs into 3D shapes and surfaces in the x-y-z coordinate system. Some of the designs build on traditional skills like draping on dress forms, and some require the invention of new processes and techniques, particularly in using new materials like filaments and powdered material.

Additionally, designers of traditional activewear need to be acquainted with *new 3D CAD tools* for exploring complex structures. Most 3D CAD tools today are developed for modeling industrial and rigid machine parts with traditional engineers in mind, not for designers who use organic forms, let alone apparel and textile designers. First, the program logic in 3D CAD modeling is dynamically different from the average 2D CAD programs. Then the program tools and even icons are often not translatable to garment design terms and tools. For instance, stitching pieces together requires the use of the "join" tool in 3D CAD modeling, and separating or cutting pieces may involve the "explode" tool. Therefore, in order to adapt to these kinds of additive technologies, designers who create conventional types of activewear need to first determine the type of 3DP that they will use, that is, whether they will use fully or partially printed components, and then, identify the key functions of the garment and select the most appropriate materials and printing method that will result in optimal perceived comfort of the wearer. Currently, the most popular and useful programs include Rhinoceros and Fusion 360 by Autodesk, and the more complex interfaces include 3Ds Max, Maya, and Blender but these are seldom used in apparel related developments. In computational 3D modeling, often parametric methods and the more innovative generative design interfaces are applied. Current programs and software plug-ins include Grasshopper, Dynamo, and Python. For 3DP, files are mostly saved in an .stl file format and often modified or revised for accuracy before printed in more straightforward 3D CAD programs, such as Netfabb and Meshmixer.

Last, *material performance and evaluation* are other critical factors of the potential obstacles in using 3DP technology for activewear. Although this chapter has mainly discussed two types of materials, nylon and TPU, there are many other points of evaluation and modification from the many properties and performance possibilities of materials for specific designs. Inevitably, the designing, sampling, and post-finishing of materials or 3D printed "textiles", as well as modifying 3D models, have become vital in the design workflow. As mentioned earlier, novel materials and printing techniques are being discovered and validated for activewear applications. There is ultimately the need for more divergent thinking of modern designers to address the potential challenges in innovative ways. In doing so, the skill sets and knowledge of the designers will be enhanced from only using conventional design and product development skills to those that are engineer-based (Sun & Zhao, 2018). In hindsight, the industry will probably expect modern designers to actively use the different technological resources in a co-designing or collaborative working environment (Sun & Zhao, 2018).

7.5 Conclusion

In this chapter, the latest developments in integrating 3DP in activewear have been given an introduction by presenting cases of multidisciplinary-based apparel and footwear designs. Three key 3DP technologies, related types of materials, and basic post-finishing techniques are explained as well. It is essential for future designers, manufacturers, and small enterprises to understand and evaluate the critical aspects of applying 3DP in apparel and footwear for ready-to-wear qualities. The key factors to consider when selecting 3DP methods and materials include: (1) wearability of the intended apparel functions and aesthetics, (2) feasibility in final construction or assembly, (3) level of difficulty in post-finishing techniques, and (4) practicality and novelty in product end-use.

In order to adapt to the new fabrication technologies and 3D CAD modeling tools, some of the apparent knowledge, skill sets, and challenges are highlighted in this chapter. It is very crucial for future designers and relevant stakeholders to address the steep learning curve of apparel practitioners who are trained to use conventional methods in their familiarization process with 3DP technologies. The highly interdisciplinary nature and application of such technologies render education or training through collaboration or co-designing inevitable and ever more invaluable. The recommendation is that designers in training work closely with material developers and 3D CAD experts to learn to use the new technologies for designing activewear. At the same time, the potential evolution of the entire activewear supply chain, as a result of novel technology applications, will lead to new industry roles, practices, and business models in the digital era.

References

Armstrong, C. (2019). Post processing and finishing. Retrieved from https://www.3dhubs.com/knowledge-base/post-processing-fdm-printed-parts.

Technical Research Centre of Finland (VTT) (2015, June 4). Cellulose turning into a super-material of the future: Broad-based cooperation multiplying the value of Finnish wood. Retrieved from https://www.sciencedaily.com/releases/2015/06/150604084445.htm.

Continuum Fashion (n.d.). Retrieved from http://continuumfashion.com/N12.php.

Cui, T. & Sun, L. (2018). Hybrid comfort: 3D printing interwoven. International Textiles and Apparel Association (ITAA) Annual Conference Proceedings. vol. 25.

Cui, T. (2019). Disruption in digital fabrication: Exploring FDM 3D printing and 3D CAD for a wearable apparel product (Doctoral dissertation). Retrieved from https://etd.auburn.edu/handle/10415/6711.

Nervous System Inc. (2014). Kinematics fold simulation. Retrieved from https://n-e-r-v-o-u-s.com/projects/albums/kinematics-fold/.

Delamore, P. (2004). 3D printed textiles and personalized clothing. Retrieved from 234 http://www.academia.edu/917613/3D_Printed_Textiles_and_Personalised_Clothing.

Jamie, D. (2017). 3D printing in sport: Feature story. Retrieved from https://www.3dnatives.com/en/3d-printing-sport-201220174/.

Lipson, H., & Kurman, M. (2013). *Fabricated: The new world of 3D printing*. Indianapolis: John Wiley & Sons, Inc.

N12: 3D printed bikini. (n.d.). Retrieved from https://www.shapeways.com/n12_bikini.

DANIT PELEG (2018). Retrieved from https://danitpeleg.com/.

Sisson, A., & Thompson, S. (2012). Three-dimensional policy: Why Britain needs a policy-framework for 3D printing. Retrieved from http://www.biginnovationcentre.com/Assets/Docs/Reports/3D%20printing%20pa.

Sun, L., & Zhao, L. (2017). Envisioning the 3D printing era: A conceptual model for the fashion industry. *Fashion and Textiles Journal*, *4*(25), 1–16.

Sun, L., & Zhao, L. (2018). Technology disruption: Exploring the changing roles of designers, makers, and users in the fashion industry. *International Journal of Fashion Design, Technology & Education,* pp. 362–374. DOI:10.1080/17543266.2018.1448462.

Tucker, E. (2015, November 4). MIT Media Lab's Biologic material opens and closes in response to humidity. Retrieved from https://www.dezeen.com/2015/11/04/mit-media-lab-tangible-media-group-biologic-material-bacteria-fashion-design/.

Varotsis, A. B. (2019). Manufacturing process explained. Retrieved from https://www.3dhubs.com/knowledge-base/introduction-material-jetting-3d-printing.

XYZ Shoe (2013). Retrieved from http://cargocollective.com/earlstewart/XYZ-SHOE.

Activewear design for competition: case study for Hong Kong rowing team

8

Chu-po Ho, Vannesa Chu
Institute of Textiles and Clothing, The Hong Kong Polytechnic University, Hong Kong, China

8.1 Introduction

High performance sportswear that incorporates innovative technologies can enhance the performance of athletes during professional competitions. For example, rowing is a seated sport. Nevertheless, the rowers still need to use all of their muscles while rowing. Their back muscles gently contract, and all four limbs are extensively stretched out. During rowing, the arms are pulling, legs are pushing, back is contracting, hips are extending, and shoulders are flexing with repeated rowing. However, function has been a priority in rowing suit designs to enhance the performance of professional rowers. This chapter provides an overview of the considerations for developing functional sports clothing for professional athletes and explains the importance of the relationship between garment fit and body movement. Other considerations include the type of material, level of comfort, and accuracy of the fit. A case study illustrates the steps that the research team took to develop a new suit for the Hong Kong rowing team for the 2014 Asian Games in Incheon. Wear trials were conducted to evaluate the performance of the new prototype.

8.2 Specific requirements of rowing as a sport

Rowing is a sport that requires the participants to use a fixed oar or a pair of oars to propel a boat. In 1896, rowing became an Olympic sport, and female rowers became involved at the Montréal 1976 Olympic Games (International Olympic Committee, 2015). Rowing has been an official event at the Asian Games since 1982 (Gagoc, 2010). Sculling and sweep-oar rowing are two typical types of competitive rowing. In sculling, the rowers use two oars. This type of rowing can be done in singles, doubles or quads. In sweep-oar rowing, the rowers hold one oar with both hands. The rowing can be done in pairs, quads or eights. The events are further divided based on the gender, age, or weight of the crew (Fédération Internationale des Sociétés d'Aviron, 2017). Lightweight events were introduced in the Atlanta 1996 Olympic Games. Before each lightweight event on each day of the competition, lightweight rowers are weighed wearing only their uniform on a test scale. The crew is disqualified from the competition if an athlete, or the crew fails to meet the required weight standards. The criteria for lightweight rowing events are as follows (Fédération Internationale des Sociétés d'Aviron, 2017):

Latest Material and Technological Developments for Activewear. http://dx.doi.org/10.1016/B978-0-12-819492-8.00008-9

- Maximum weight of an individual lightweight male rower or a lightweight male single sculler—72.5 kg,
- Maximum average weight of a lightweight men's crew—70 kg,
- Maximum weight of a lightweight female rower or a lightweight female single sculler—59 kg, and
- Maximum average weight of a lightweight women's crew—57 kg.

8.2.1 Biomechanics of rowing

The goal of rowing is to ensure that the rowers and the boat cover the required distance as fast as possible from start to finish of the race. Rowing is a competitive sport that requires the rower to have a high level of energy and a high muscle mass to overcome the wind and drag resistance from the forward motion of the boat as well as to balance the vessel (di Prampero, Cortili, Celentano, & Cerretelli, 1971; Secher, 1990). Therefore, the mechanics of rowing differ from most other types of sports. The emphasis is on both arm and leg movements, with the two legs working in the same phase while the body is supported by a seat. The cycle of a rowing stroke involves four phases: catch, drive, extraction, and release phases. First, there is the catch, in which the oar is dipped into the water, following which the rower puts pressure onto the oar to move the boat in the drive phase. When the oar is extracted from the water, this is called the release phase. Finally, the rower prepares for the next stroke, which is the recovery phase. During competitions, rowers repeat these phases as precisely as possible for an optimal rowing performance (Whitwell, n.d.). Smith and Loschner (2002) stated that a skillful rower should maintain good stroke-to-stroke consistency.

The distance of a standard international race is 2000 m straight for men and 1000 m straight for women, and each race generally lasts six to eight minutes (Smith & Hopkins, 2012). However, rowing is affected by environmental factors, such as cold weather, rain and wind (de Campos Mello, de Moraes Bertuzzi, Grangeiro, & Franchini, 2009). It is a demanding sport, in which the athletes must be suitably tall and they must not only have great physical strength, high levels of aerobic fitness and anaerobic capacity, but also possess strong core balance and cardiovascular endurance (Hagerman, Hagerman, & Mickelson, 1979; Shephard, 1998; Wright, Nicoletti, & Shephard, 1977).

8.2.2 Design requirements of rowing uniform

Rowing is a demanding sport, so the requirements of the rowing uniform are equally as demanding. The design of the rowing shirt must comply with rules. According to FISA (2017), the racing uniform must be a shirt and shorts or a one-piece suit and it must be the same for all of the team members. The uniform should not add to the cost of taking part in the sport and everyone on the team should have access to the uniform. The clothing also should not change the nature of the sport; provides no advantage to other team members; is environmentally harmless and sound; and contributes to the positive development of rowing by maintaining the principles of the sport.

There have been reports that the body dimensions of national-level rowers have increased over time (Carter, 1982a, 1982b). Conventional approaches to producing the uniforms of rowers in terms of sizing and fitting can therefore no longer meet the functional demands of rowers. In addition, weight is also a concern even though the biomechanical and physiological characteristics of rowers are usually regarded as the most important determinants of success, which is particularly true for lightweight rowers. Therefore, the weight of the fabric used for their uniform is one of the key considerations in the functional design of rowing shirts. The weight of the fabric used can alleviate the concerns around weight from the heavy material used so that rowers can consume relatively more food to have more energy for rowing. Furthermore, it is important to examine recent issues related to the fit and perceived comfort of rowing shirts as well as the amount of body movement allowed. Engineered garment designs can therefore provide competitive advantages that overcome the limitations of the biomechanics and physiology of rowing athletes.

Many factors need to be considered to improve the performance of athletes so that they put forth their best in any sporting activity. Three major factors that contribute to either the success or failure of an athlete are: (a) her/his ability, (b) the equipment and facilities, and (c) the engineered clothing (Chowdhury, Alam, Mainwaring, Beneyto-Ferre, & Tate, 2012; Yan, Yi, Tao, Lin, & Li, 2011). In particular, research has indicated that functional sportswear will improve athletic performance after studying the role of engineered clothing on the human body.

The comfort of clothing is affected by the environment, moisture and thermal management, fabric handle, psychologically perceived comfort and tested results (Thiry, 2003). The comfort of athletic clothing depends on the type of fiber(s) used, spinning method, weaving or knitting parameters, fabric weight and density, functional finishing, fit, and manufacture (Ishiaque, 2001; Pontrelli, 1990; Smith, 1986; Sontag, 1985). Wear comfort is an important quality of sportswear because it affects performance and efficiency (Bartels, 2005; Gill & Prendergast, 2015). The human body performs at its maximum efficiency when comfort conditions are met (Choudhury, Majumdar, & Datta, 2011). Mecheels (1998) defined four aspects of wear comfort:

(1) *Thermophysiological comfort,* which is characterized by breathability, thermal insulation and moisture management.
(2) *Ergonomic comfort,* which is related to the fit and the freedom of movement of a garment, and is determined by the pattern of the garment and the elasticity of the fabrics used.
(3) *Skin sensorial comfort,* which mainly depends on the perceived physical sensations of the wearer toward a textile that is in direct contact with the skin.
(4) *Psychological comfort,* which is determined by personal preferences and fashion. The level of psychological comfort is subjectively perceived through thermal, visual and tactile sensations, body–garment interactions and external environmental effects (Choudhury et al., 2011; Li, 2001).

Therefore, wear comfort can be measured by conducting wear trials with human subjects, either in a climatic chamber or under practical conditions (Bartels, 2005).

8.3 Design and development of new uniform for Hong Kong rowing team

In light of the importance of the uniform of rowing athletes, the Institute of Textiles and Clothing of the Hong Kong Polytechnic University collaborated with the Hong Kong rowing team to develop a racing suit. The project included tasks such as material sourcing, fabric testing, prototype development, a wear trial, and prototype evaluation. Therefore, a detailed plan was drafted before the commencement of the project. The tasks, contents, timeline and objectives are listed in Table 8.1. All of the involved parties, including the research team of the Institute of Clothing and Textiles, manufacturer, and the rowing team, met once to discuss the plan and agree on the timeline.

8.3.1 Size and fit

The fit of sports clothing is of paramount importance, because it is directly related to the performance of the athlete and the outcome of the competition (Gill & Prendergast, 2015). Well-fitting and -designed clothing should not restrict body

Table 8.1 Tasks and objectives of development of new racing suit for Hong Kong rowing team.

Task	Details	Objectives
1	Taking measurements of each rower, and conducting focus group interview	• To measure static and sculling movements • To identify individual attributes of current racing suit • To identify problems of current racing suit
2	Material sourcing, pattern-making, and suit-making	• To brainstorm on potential approaches for solving issues • To develop first prototype
3	Fitting (Prototypes fabricated based on body measurements and individual fit preferences.)	• To allow rowers the opportunity to don their new prototype and then undergo a one-hour training session in an indoor training center. • To gather feedback on the fabrication, size, fit, cut and seam details of the initial prototypes.
4	First wear trial (Prototypes fabricated based on feedback from initial fitting.)	• To allow rowers to don modified prototype for daily training. • To gather feedback four weeks after wear trial.
5	Second wear trial (Prototypes fabricated based on feedback from first wear trial.)	• To allow rowers to don their modified prototype for training and competitions in Qingdao, Lucerne and Tilburg. • To gather feedback on modified prototypes via online survey.
6	Production of new racing suit ready for use at 2014 Asian Games	• Manufacturers to produce new racing suit for 2014 Asian Games after evaluating survey response.

movement for the occasion in which the garment is worn (Ashdown, 2011). In general, studies have concluded that determining the potential fit of a garment on live models makes sense (Wibowo et al., 2018) because real body shapes and movement are involved (Fan, Yu, & Hunter, 2004). Therefore, perception of the wearer is a key consideration in evaluating engineered garment fit during the design process.

Market research has also shown that sports participants from 10 different countries have considered the following to be significant features of clothing for sports activities: comfort, freedom of movement, breathability, machine washability, durability, fit, and light in weight (Voyce, Dafniotos, & Towlson, 2005). Among these features, freedom of movement is directly related to the fit of a garment.

In terms of fit, garment ease is the difference in the distance between a garment in a given area and the body itself (Alexander, Connell, & Presley, 2005) so that the wearer feels comfortable when wearing the garment. During pattern-making, ease can be integrated by increasing the extra space (Geršak, 2014) that allows the body to move within the garment. Factors that include the design, type of garment and body, fabric, garment function, and personal preferences are considered in determining the ease allowance (Cooklin, 1997). Generally, there are two different types of ease: wearing ease and design ease (Fan et al., 2004; Geršak, 2014). Wearing ease is the space required between the garment and the body for comfort and wearability. This enables the wearer to perform a range of movements, such as stretching, without the clothing being pulled and strained beyond the natural relaxed position (Liechty, Pottberg, & Rasband, 1992). Design ease is the space added for personal preference and aesthetic purposes (Geršak, 2014). This type of ease in a pattern is not standardized and changes depending on the fashion trends.

There are four types of ease allowance which contributes to wearability and comfort: standard, dynamic, mechanical, and ergonomic (Chen, Zeng, Happiette, & Bruniaux, 2008; Geršak 2001; Geršak & Marčič, 2013; Ng, Liu, & Yu, 2007). They are defined as follows.

(1) *Standard ease* allowance is the difference between the maximum and minimum perimeters of the body based on the standard body shape in a standing or sitting position.
(2) *Dynamic ease* allowance is the distance between the clothing and the wearer that allows the wearer to perform certain extreme movements and postures or accommodates different body shapes.
(3) *Mechanical ease* allowance is related to the mechanical properties of the fabric that contribute to perceived comfort, such as the tensile and shear modulus.
(4) *Ergonomic ease* allowance is based on an increase in body dimensions when an individual is in motion (Assembil, 2013). This allowance of ease involves adjusting the design and fit, and varying the fabric, structure and construction of a garment to enable unrestricted movement that would meet the demands of different targeted users.

It is important that garment ease is optimized for fit and wearability (Fan et al., 2004; Geršak, 2014; Liechty et al., 1992). It is also important to consider the specific functions of a sports garment and its purpose when creating a design to maximize the performance of the wearer. For instance, the anatomy of a human (Watkins, 1995; Ashdown, 2011) has been investigated in different studies to obtain the functional considerations around garment ease for dynamic positions (Choi & Ashdown, 2011).

A garment with an engineered fit is defined as a clothing item that fulfills a functional need. For instance, Kirk and Ibrahim (1966) concluded that garment fit is an essential factor in meeting the skin strain requirements based on the clothing design and the ratio of the clothing size to body size. Mass-produced garments offer an approximate fit while made-to-measure garments have a better fit because the base pattern is adjusted based on the body dimensions of an individual (Watkins, 2011). However, the body shape of elite athletes may substantially vary from the norms in recent sizing and garment development practices, which may affect their performance during sports participation (Gill, 2015). Moreover, the fitting methods that are used for mass produced clothing patterns are subjected to the limitations of currently available technology. Therefore, three-dimensional (3D) scanning of the body is a much more appropriate method for an engineered fit, which is used in this study.

The specific movements of rowing require the uniform worn during this activity to have special characteristics. That is, rowers scull by controlling their oars and seat. To finish the cycle of sculling, each rower pulls the oars back to push the water. In doing so, the rower pulls the handles of the oars towards her/his abdomen. If her/his racing suit is too loose, the excess drape may interfere with the handles of the oars and affect the rhythm of the rowing. While a tighter or more fitted racing suit may prevent this problem, an overly tight suit not only affects the comfort of the rower, but also limits the extent of his/her ability to stretch during the rowing movement (Ho & Au, 2016).

Therefore, to address these issues, this project involved fifteen rowers with nine males and six females on the Hong Kong rowing team. During the first meeting, the rowers commented on their current racing suit, which was ordered from an overseas company and available in a small, medium or large size. As the rowers all have a different body size, the standardized suits could not accommodate their individual fit. Like other commercialized garments, this sizing system cannot tailor to every body size. Therefore, a made-to-measure approach was adopted in the project to prevent the potential problems caused by a poor fit and accommodate the different body shape and size of the rowers. This approach is the key to improving the racing suit. To do so, a 3D body scanner (ScanWorX) that is available at the Hong Kong Polytechnic University was used to obtain quick but accurate body measurements of the rowers.

Table 8.2 is a summary of the body measurements. The difference in body height between the shortest and tallest male rowers is substantial at nearly 21 cm as opposed to the female rowers at 9 cm. However, there are greater differences in the chest, waist, and hip measurements for the female rowers than the male rowers.

Although the 3D body scanner can accurately and conveniently capture the body measurements when the rower is standing in a static position, and the amount of extension of the body during rowing, the research team further solicited the assistance of a skilled fitting technician as scanning would not satisfy the individual expectations of the garment fit. The research team expected that the rowers could perform rowing motions while wearing the first prototype so that the technician could observe the amount of ease. Moreover, the rowers and technician could better present the expectations and preferences around individual fit so that the research team could immediately make them on the prototypes. Thus, 3D scanning provided a quick and reliable way of taking body measurements during the early stages of the prototype

Table 8.2 Body measurements of rowers.

	Range of body measurements (cm)			
	Height	**Chest**	**Waist**	**Hip**
Males	166.5–187.4	43–48	93–99	79–89
Females	162.5–171.5	86–93	74–83	87–98

development, while the fitting technician refined the measurements and pattern as well as addressed the subjective fit of the rowers. Finally, the research team produced 15 different prototypes for the rowing team, which matched the size and fit specifically of each rower on the team.

8.3.2 Material

Fabrics are the basic structure of garments and have a direct bearing on how fit is perceived with body movement. Huck (1988) investigated garment restrictions of body movement due to fabric characteristics, while Kirk and Ibrahim (1966) concluded on the need for elasticity and recovery in functional fabrics. However, neither study specifically focused on sportswear. Although it is difficult to evaluate in practice, fabric characteristics including stretch and elasticity are important for retaining garment function during movement (Bartels, 2005; Kirk & Ibrahim, 1966; Watkins, 1995). For instance, Watkins (1977) reported that the type of fabric used for the uniform of ice hockey players needs to provide the necessary stretch for all of their movements on the ice. Denton (1971) categorized three types of stretch in clothing: (a) comfort stretch, in which the clothing stretches with movement, but is not too tight fitting; (b) stretch-to-fit, in which the garment is a close fit but does not apply substantial pressure onto the body; and (c) power stretch, which exerts compressive pressure onto the body (Ashdown, 2011).

The use of fabrics with optimal stretch and recovery has been a major design goal to ensure the functionality of active garments. Properties such as fit, comfort, freedom of movement and shape retention after repeated washings are determinants of the quality of such garments (Voyce et al., 2005). Knitted fabrics are commonly preferred in sportswear applications because they are stretchable and provide a snug fit (Mikučionień & Milašiene, 2013; Sabir & Wood, 2015). Knits were introduced to the sportswear market in the early 1980s (Abd El-Hady & Abd El-Baky, 2011). The looping structure gives knitted fabrics an inherent stretch so that they can recover their shape when tension is released (Ashdown, 2011). The high stretchability of knitted fabrics means that they are a common material used for all kinds of sportswear, because they provide flexibility to enhance movement (Behera, Mani, Mondal, & Sharma, 2002; Buhler & Iyer, 1988). Therefore, knitted garments move with the body without inhibiting the performance of the wearer. One of the most popular knitted fabrics is Lycra. Lycra, the first human-made elastomeric yarn (spandex), was invented in 1958. This knit can stretch four to seven times its initial length but also recover (Voyce et al., 2005). The elastomericity of Lycra has resulted in its popularity in active wear (Singha, 2012) because the material facilitates body movement with ease

Table 8.3 Test results for fabric of new racing suit.

Test	Standard	Details/rating
Air permeability (after 5 cycles of washing)	ISO 9237: 1995	22 (cc/sec/cm^2 at 100 Pa)
Water vapor permeability (after 5 cycles of washing)	ASTM D96/E96M: 2013	1200 (g/m^2)
Bursting strength (psi)	ASTM D3786: 2006	48 psi
Abrasion resistance	ASTM D4966: 1988	No breaking after 10,000 rubs
UV resistance (UPD rating)	AS/NZS 4399: 1996	50+

and does not inhibit performance. A small amount of Lycra or just 2% is sufficient to improve the shape retention of the garment and improve overall movement of the wearer (Voyce et al., 2005).

The fabric of the new racing suit in the project of interest here is a 22 gauge interlock knitted fabric with a 92% polyester and 8% Lycra blend to enhance stretching. The fabric weight is 200 g/m^2, which is 40 g/m^2 lighter than that of the previous suit. Both polyester and Lycra are synthetic materials. Synthetic fibers are commonly used for sportswear because they provide a good balance of softness, are light in weight, and those structurally stable and can wick away moisture (Uttam, 2013).

During the first meeting with the coach and rowers of the Hong Kong rowing team, they emphasized that a uniform with an extra-smooth surface fabric is undesirable because the rowers would not be able to move on the boat seat with the right rhythm during sculling, especially when there is sea water on the seat.

Also, rowing activities take place in bodies of water and outdoors, so they expose rowers to humid environments and ultraviolet (UV) light from sunshine. Therefore the fabric of the racing suit had to include high air permeability and UV protection. In addition, their body is often wet from contact with sea water so a fabric that is quick-drying would keep the skin dry after exposure to sea water. Therefore, it was essential that the material for the racing suit could properly manage moisture to keep the rowers dry during rowing.

Before sourcing the selected fabric for the racing suit from the market, some key tests were conducted to evaluate the functionality of the fabric, such as air and water vapor permeabilities and UV resistance. Table 8.3 lists some of the tests performed and the results on the fabric used for the new racing suit (Ho & Au, 2016).

8.3.3 Design details

The seam type and pattern construction are two important details that were considered during the design process because they directly affect the scratchiness of the racing suit, especially during sculling. The new racing suit had flat seams which were finished by using a 607 coverstitch formed with 4-needle threads, a top cover thread and a bottom looper thread interlaced together (ASTM D-6193). The neckline, armhole and thigh components were finished with a 406 coverstitch (ASTM D-6193).

Figure 8.1 Design sketch of new racing suit.

Together with an appropriate sewing thread, these types of stitches reduce scratchiness (Ho & Au, 2016). Fig. 8.1 Fig. shows the prototype design sketch.

During training and competition, most of the male rowers do not wear undergarments beneath their racing suit to keep friction low when they are sitting on the seat and rowing. To account for this, the research team removed the centre front and back seams as well as the seams on the bottom part of the suit to eliminate potential scratchiness in these areas. In addition, two side-seams were moved forward to improve the aesthetics of the suit (Ho & Au, 2016) (see Fig. 8.2 Fig.).

Traditional embroidery (as shown in Fig. 8.3A and B) unavoidably exposes the sewing thread on the wrong side of the suit, which might cause friction to the skin of the chest (Ho & Au, 2016). The national flag at the front was therefore changed from an embroidered to a heat-transfer polyurethane (PU) badge (Fig. 8.3C and D) to retain the smoothness of the wrong side of the fabric. In addition, the research team observed that the current embroidered badge might pucker on the fabric which is due to the penetration of the embroidery thread (Fig. 8.3B). This puckering effect is not uncommon when embroidery is applied to a stretchable fabric. However, this penetration is not an issue with a heat-transfer PU badge (Fig. 8.3D). To address the puckering problem, the research team proposed either a screen-printed or heat-transfer 5-trial samples and cost, everyone agreed to use the heat-transfer PU badge based on the two following reasons:

(1) The white pigment on the screen print might fade after repeated launderings and tumble drying. The heat transfer PU badge would be more durable in this case.

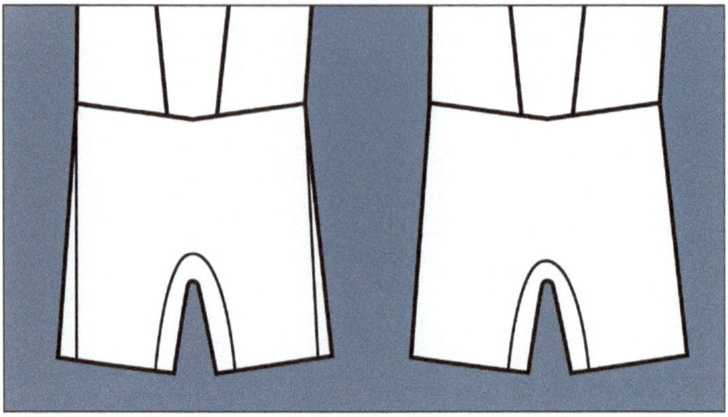

Figure 8.2 Construction of bottom part of rowing suit for male rowers. (A) Front, (B) back.

(2) The heat-transfer badge may incur a higher cost during production, but provides a more refined look to the racing suit. It is a tradition of rowers to exchange their racing suit with that of other national teams after large-scale competitions. The new racing suit could improve the image of the Hong Kong rowing team to others.

Fig. 8.4 shows the new racing suit developed for the male rowers.

8.4 Evaluation of new prototype

Before producing the final version of the racing suit, several prototypes were produced for each rower to evaluate the performance of the new design. This process was divided into fitting (one round) and wear (two rounds) trials, which were conducted at different locations.

Fitting took place at the training center. After collecting the measurements from the 3D scanner, the manufacturer produced the first prototype for each rower. Each prototype is unique in size and fit. Fig. 8.5 shows the first prototype developed for the fitting session. As the key purpose of the prototype is to provide a visual of the new racing suit and collect feedback from the rowers, the Hong Kong flag was not placed on all of the prototypes to reduce production cost and time. All of the rowers were asked to provide verbal feedback on all aspects of their prototype after they had donned the suit for their indoor training session which lasted for more than an hour.

To evaluate individual fit preferences, some of the rowers were asked to swap their suit with other rowers to compare the different fit on certain parts of the body. This was a good way for the rowers to evaluate the individual fit (Ho & Au, 2016).

The questions listed in Table 8.4 were used to guide the rowers, and the same questions were raised during the wear trials so that the research team could review and compare the data for the different prototypes.

Fig. 8.6 shows one of the rowers wearing the new prototype at the fitting session. He commented that the fit of the waist is too loose which might affect rowing during

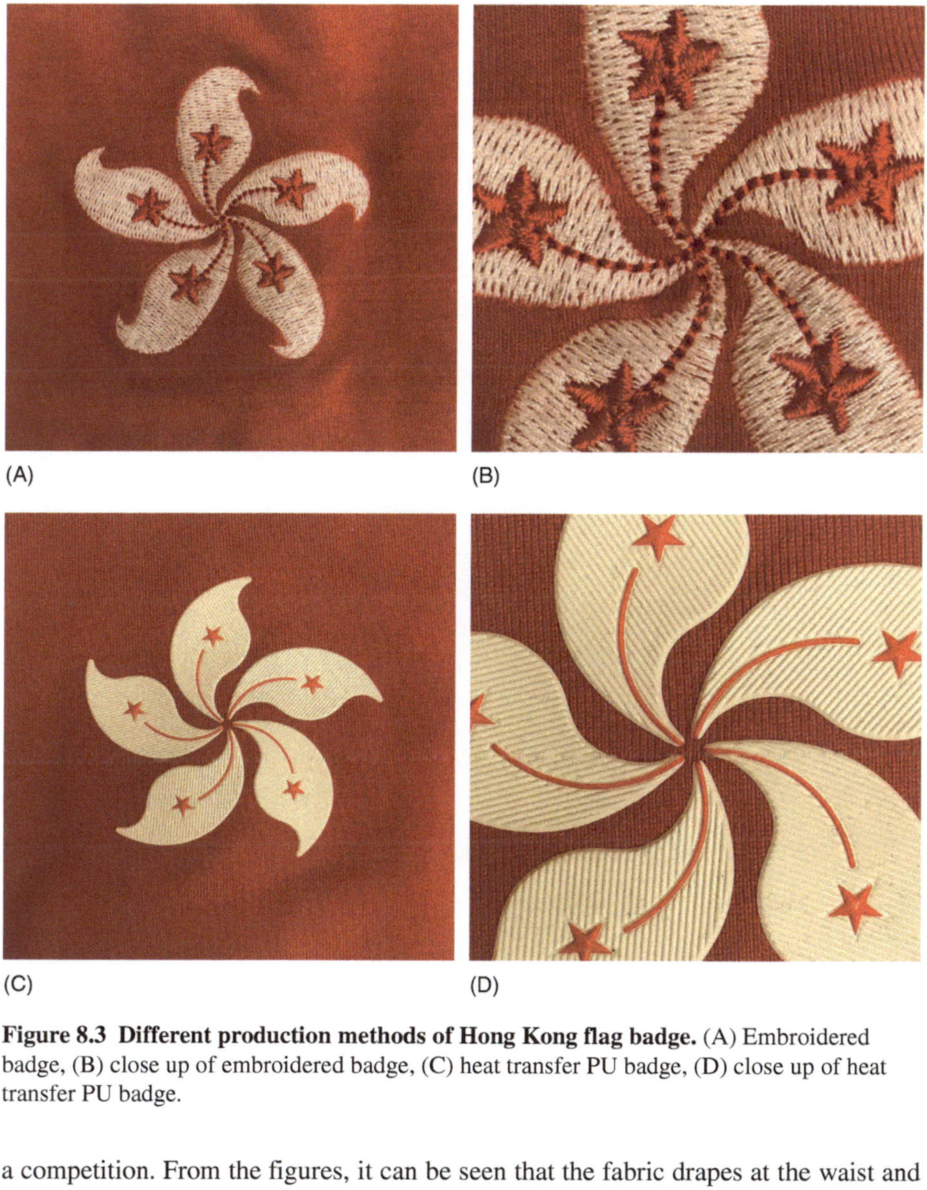

Figure 8.3 Different production methods of Hong Kong flag badge. (A) Embroidered badge, (B) close up of embroidered badge, (C) heat transfer PU badge, (D) close up of heat transfer PU badge.

a competition. From the figures, it can be seen that the fabric drapes at the waist and back area. Before the first fitting, the technician added garment ease at the back after taking the rowing movement into consideration. This approach is commonly used for non-stretchable garments to give ease of movement to the wearer. After obtaining feedback from the rowers, it was found that no garment ease is needed as the fabric has high stretchability already. However, once the garment ease was not implemented, the research team had to be sure that the waist would not be too tight for the individual rowers. Therefore, this fitting session was crucial to obtain information on the first impressions of the rowers about the fit of the new racing suit. In the following

Figure 8.4 New racing suit for male rowers.

prototype, no garment ease was applied to the waist and back. After receiving all the comments, the technician made marks on the suits directly for follow up, and then the second batch of prototypes were produced for the formal wear trials.

The wear trial consisted of two phases, which were conducted in different places. The first prototype was donned by the rowers for their daily rowing practice session in Hong Kong for 4 weeks so that they could form a deeper impression of their prototype. The feedback from this wear trial was expected to be more in-depth because the rowers wore their prototype for long hours, laundered it every day, and exposed it to outdoor conditions such as sunshine, rain, and sea water. After 4 weeks of training, the research team met with the rowers at the training center, and the rowers were invited to provide verbal feedback to the questions listed in Table 8.4.

Fig. 8.7 A–D shows the second prototype developed for the first formal wear trial. As with the suit that was developed for the first fitting, the heat-transfer badge was not applied. In this version, the back and waist area were smoother than those of the first prototype. After the first wear trial, all of the rowers agreed that extra garment

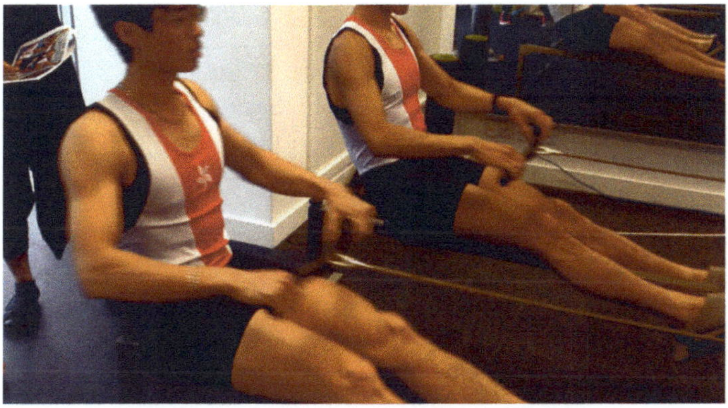

Figure 8.5 Fitting session at training center.

ease was not necessary for the back and waist of the racing suit; however, some modifications were required, primarily to the leggings of the suit. Most of the rowers preferred longer and looser leggings which are more comfortable during rowing.

After modifying the second prototypes, the third prototypes were sent to the rowers for the second round of wear trials. The rowers wore their second prototype in Qingdao, Lucerne, and Tilburg for their training sessions and regional competitions before the start of the Asian Games. Feedback was provided in an online survey that

Table 8.4 Evaluation questions.

Question	Objectives
After the training, do you think that the new suit increases your freedom of movement (including contact with the seat)? If no, can you give some comments for the following body parts: neckline, armhole, chest, waist, back, crotch and thighs?	To evaluate: (1) the smoothness and stretchability of the fabric, and (2) size and fit.
After the training session, do you think that the new suit is better comfortable on your skin? If no, can you comment based on the following body parts: neckline, armhole, chest, waist, back, crotch and thighs?	To evaluate: (1) size and fit; (2) seam and stitch type; (3) construction; and (4) softness of the fabric.
After the training session, do you think that the new suit kept your skin dry?	To evaluate the moisture management properties of the fabric
After the training session, do you think that the new suit meets your individual fit preferences?	To evaluate: (1) size and fit, and (2) smoothness of the fabrication
Do you think that the new Hong Kong badge reduces or eliminates feelings of scratchiness on your chest? (This question was not applicable in the fitting session.)	To evaluate the effectiveness of using a heat transfer badge in reducing feelings of scratchiness on the chest.
Do you think that the new suit makes your private parts itch? (Only applicable to the male rowers)	To evaluate: (1) construction of the suit, and (2) seams and stitches
What are your overall comments for the new suit?	To evaluate overall comfort

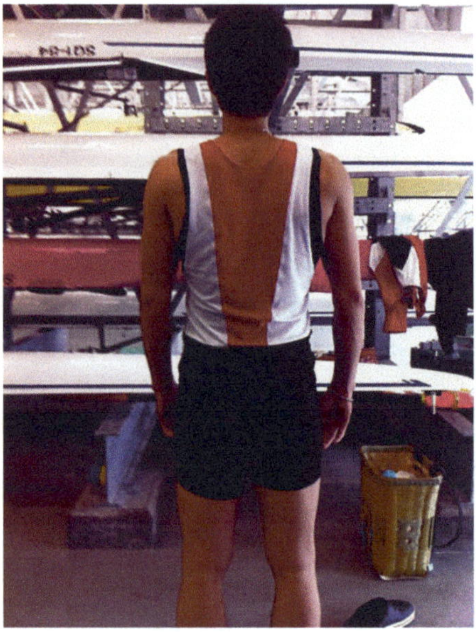

Figure 8.6 First prototype.

the rowers were asked to complete within four weeks into their training for the competition. The survey was based a 7-point Likert scale through which the rowers rated the level of perceived comfort of the second prototype, with 3 as the highest perceived comfort and -3 the lowest perceived comfort, and 0 is neutral. An open-ended question on the overall perceived comfort was available at the end of the survey so that the rowers could provide input on any issues about the second prototype. Any part of the survey with a mean less than 0 would be revisited by the research team and the rowers. Table 8.5 summarizes the main comments of the rowers.

The rowers rated all of the factors with a 2 or 3, and no negative comments related to the new prototype were provided. After considering all of the comments, the racing suit was formally produced. Fig. 8.8 Fig. shows the new racing suit worn by one of the male rowers who is the first in Hong Kong to receive a gold medal at the Asian Games for rowing.

Table 8.5 Comments of rowers from online survey.

Comments
• The new prototype fits my body. I can move freely when rowing. • The bottom part of the new prototype is more comfortable. It reduces the friction to my private parts (this comment was made by a male rower) • The lighter weight of the suit means that I can eat a bit more so that I have more energy. • This is even more important for the lightweight competition.

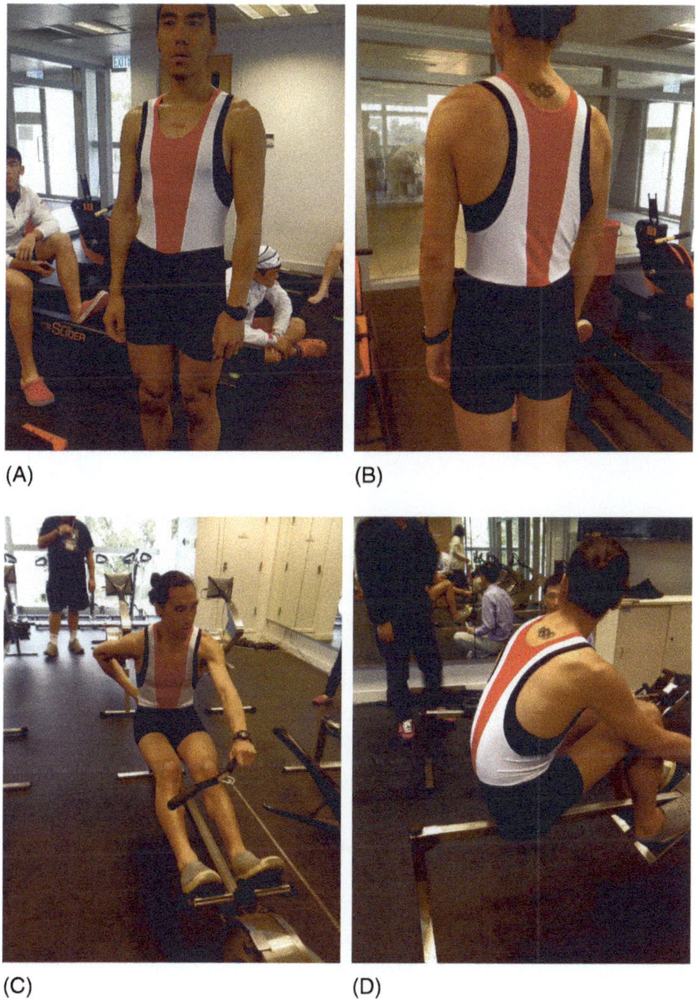

Figure 8.7 Second prototype in first wear trial. (A) Front view, (B) back view, (C) front view during training, (D) side view during training.

8.5 Conclusion

Producing activewear that is light in weight and comfortable to wear is not an easy task. The researchers had to develop a set of fabrication testing standards based on the special requirements of a racing suit for rowing, including one that offers UV protection, manages moisture, is light in weight, soft and durable, and has good color fastness.

It has been shown that fabric and fit are the two major considerations for the development of a new racing suit for the Hong Kong rowing team. The requirements for the fabric of the suit are that it is stretchable, permeable, abrasion-resistant, hygroscopic,

Figure 8.8 First gold medal for rowing at Asian Games.

allows heat to pass through, wicks away sweat and UV-resistant, so as to provide athletes with a next-to-skin and comfortable fit. During the development of the rowing suit, the weight, cut and other properties of the fabric all catered to the needs of the athletes.

The rowers must stretch their entire back, arms, and legs to perform the sculling movement unique to rowing. Therefore, the racing suit had to accommodate these specific body movements. Other aspects such as the seams and the flag on the suit, as well as the ease are also vital and modified to accommodate individual fit.

To help the rowers do well in competitions, the research team at the Institute of Textiles and Clothing of the Hong Kong Polytechnic University has successfully developed new racing suits that are unique to each rower. Ultimately, the rowers wore their new suit at the 2014 Asian Games which ultimately resulted in the first gold medal and four silver medals for the Hong Kong rowing team.

References

Abd El-Hady, R. A. M., & Abd El-Baky, R. A. A. (2011). Enhancing the functional properties of sportswear fabric based carbon fibre. *Asian Journal of Textiles*, *1*(1), 14–26.

Alexander, M., Connell, L. J., & Presley, A. B. (2005). Clothing fit preferences of young female adult consumers. *International Journal of Clothing Science and Technology*, *17*(1), 52–63.

Assembil Books. (2013). *How patterns work: The fundamental principles of pattern making and sewing in fashion design*. London: Assembil Books.

Ashdown, S. P. (2011). Improving body movement comfort in apparel. In G. Song (Ed.), *Improving comfort in clothing* (pp. 278–302). Cambridge: Woodhead Publishing Limited.

Bartels, V. T. (2005). Physiological comfort of sportswear. In R. Shishoo (Ed.), *Textiles in Sport* (pp. 177–203). Cambridge: Woodhead Publishing Limited.

Behera, B. K., Mani, M. P., Mondal, A. K., & Sharma, N. (2002). Comfort behaviour of cotton polypropylene based bi-layer knitted fabrics. *Asian Textile Journal*, *8*, 61–67.

Buhler, M., & Iyer, C. (1988). An investigation into opportunities for further developments of functional knitwear for the sportswear sector. *Knitting Technique*, *10*(5), 303–307.

Carter, J. E. L. (Ed.). (1982a). *Physical structure of olympic athletes: Part I. Body composition of athletes*. Basel: Karger.

Carter, J. E. L. (Ed.). (1982b). *Physical structure of olympic athletes. Part II. Age and body size*. Basel: Karger.

Chen, Y., Zeng, X., Happiette, M., & Bruniaux, P. (2008). A new method of ease allowance generation for personalization of garment design. *International Journal of Clothing Science and Technology*, *20*(3), 161–173.

Choi, S., & Ashdown, S. (2011). 3D body scan analysis of dimensional change in lower body measurements for active body positions. *Textile Research Journal*, *81*(1), 81–93.

Choudhury, A. R., Majumdar, P. K., & Datta, C. (2011). Factors affecting comfort: Human physiology and the role of clothing. In G. Song (Ed.), *Improving comfort in clothing* (pp. 3–60). Cambridge: Woodhead Publishing Limited.

Chowdhury, H., Alam, F., Mainwaring, D., Beneyto-Ferre, J., & Tate, M. (2012). Rapid prototyping of high performance sportswear. *Procedia Engineering*, *34*, 38–43.

Cooklin, G. (1997). *Garment technology for fashion designers*. Oxford: Blackwell Science.

de Campos Mello, F., de Moraes Bertuzzi, R. C., Grangeiro, P. M., & Franchini, E. (2009). Energy systems contributions in 2000 m race simulation: a comparison among rowing ergometers and water. *European Journal of Applied Physiology*, *107*(5), 615.

di Prampero, P. E., Cortili, G., Celentano, F., & Cerretelli, P. (1971). Physiological aspects of rowing. *Journal of Applied Physiology*, *31*, 853–857.

Denton, M. J. (1971). *Fit, stretch, and comfort. Paper presented at the Third Shirley International Seminar: Textiles for Comfort*. Manchester, England: The Cotton Silk and Man-made Fibres Research Association, Shirley Institute, , pp. 1–11.

Fan, J., Yu, W., & Hunter, L. (2004). *Clothing appearance and fit: Science and technology*. Cambridge: Woodhead Publishing Ltd.

Fédération Internationale des Sociétés d'Aviron (2017). FISA Rule Book. Retrieved from http://www.worldrowing.com/mm//Document/General/General/13/08/90/FISArulebookEN-2018finalweb3_Neutral.pdf.

Gagoc (2010). Rowing. Retrieved from http://www.china.org.cn/sports/2010asiangames/2010-10/12/content_21104888.htm.

Geršak, J. (2014). Wearing comfort using body motion analysis. In D. Gupta, & N. Zakaria (Eds.), *Anthropometry, Apparel sizing and design* (pp. 320–333). Cambridge: Woodhead Publishing Ltd.

Geršak, J. (2001). Designing a Garment System from the Point of View of Thermophysiological Comfort. *Annals of DAAAM for 2001 & Proceedings of the 12th International DAAAM Symposium 'Intelligent Manufacturing & Automation: Focus on Precision Engineering'*. Jena University of Applied Sciences, Jena, Germany, Vienna, DAAAM International, pp. 157–158.

Geršak, J., & Marčič, M. (2013). The complex design concept for functional protective clothing. *Tekstil: Journal of Textile & Clothing Technology*, *62*(1/2), 38–44.

Gill, S. (2015). Body scanning and its influence on garment development. In S. G. Hayes, & P. Venkatraman (Eds.), *Materials and technology for sportswear and performance apparel* (pp. 322–337). Boca Raton, FL: CRC Press.

Gill, S., & Prendergast, J. (2015). Garment fit and consumer perception of sportswear. In S. G. Hayes, & P. Venkatraman (Eds.), *Materials and Technology for Sportswear and Performance Apparel* (pp. 245–260). Boca Raton, FL: CRC Press.

Hagerman, F. C., Hagerman, G. R., & Mickelson, T. C. (1979). Physiological profiles of elite rowers. *The Physician and Sportsmedicine*, *7*(7), 74–83.

Ho, C., & Au, Y. (2016). Development of functional racing singlet for professional rowers. *International Journal of Fashion Design, Technology and Education*, *10*(2), 137–145.

Huck, J. (1988). Protective clothing systems: A technique for evaluating restriction of wearer mobility. *Applied Ergonomics*, *19*(3), 185–190.

International Olympic Committee (2015). Rowing: History of rowing at the Olympic Games. Retrieved from https://stillmed.olympic.org/AssetsDocs/OSC%20Section/pdf/QR_sports_summer/Sports_olympiques_aviron%20_eng.pdf.

Ishiaque, S. M. (2001). Engineering comfort. *Asian Textile Journal*, *10*(11), 36–39.

Kirk, W., & Ibrahim, S. (1966). Fundamental relationship of fabric extensibility to anthropometric requirements and garment performance. *Textile Research Journal*, *36*(1), 37–47.

Li, Y. (2001). The science of clothing comfort. *Textile Progress*, *31*(1/2), 1–135.

Liechty, E., Pottberg, D., & Rasband, J. (1992). *Fitting & pattern alteration: A multimethod approach*. New York: Fairchild Publications.

Mecheels, J. (1998). *KoÈrper-Klima-Kleidung: Wie funktioniert unsere Kleidung?* Berlin: Schiele & SchoÈn.

Mikučionienė, D., & Milašiene, D. (2013). The influence of knitting structure on heating and cooling dynamic. *Materials Science*, *19*(2), 174–177.

Ng, R., Liu, W., & Yu, W. (2007). Dynamic stress distribution on garment. *Tekstil*, *56*(2), 123–127.

Pontrelli, G. J. (1990). Comfort by design. *Textile Asia*, *21*(1), 52–61.

Sabir, T., & Wood, J. (2015). Fabrics for performance clothing. In S. G. Hayes, & P. Venkatraman (Eds.), *Materials and Technology for Sportswear and Performance Apparel* (pp. 87–99). Boca Raton, FL: CRC Press.

Secher, N. H. (1990). Rowing. In T. Reilly, N. Secher, P. Snell, & C. Williams (Eds.), *Physiology of sports* (pp. 259–286). London: E & F.N. Spon.

Shephard, R. J. (1998). Science and medicine of rowing: A review. *Journal of Sports Sciences*, *16*(7), 603–620.

Singha, K. (2012). Analysis of spandex/cotton elastomeric properties: Spinning and applications. *Analysis*, *2*(2), 11–16.

Smith, J. E. (1986). The comfort of clothing. *Textiles*, *151*, 23–27.

Smith, R., & Loschner, C. (2002). Biomechanic feedback for rowing. *Journal of Sports Sciences*, *20*(10), 783–791.

Smith, T. B., & Hopkins, W. G. (2012). Measures of rowing performance. *Sports Medicine*, *42*(4), 343–358.

Sontag, M. S. (1985). Comfort dimensions of actual and ideal insulative clothing for older women. *Clothing and Textiles Research Journal*, *4*(1), 9–17.

Thiry, M. C. (2003). Feelin' fine: Textiles bring a sense of comfort. *American Association of Textile Chemists and Colorists*, *3*(3), 9–14.

Uttam, D. (2013). Active sportswear fabrics. *International Journal of IT, Engineering and Applied Sciences Research*, *2*(1), 34–40.

Voyce, J., Dafniotos, P., & Towlson, S. (2005). Elastic textiles. In R. Shishoo (Ed.), *Textiles in sport* (pp. 204–230). Cambridge: Woodhead Publishing limited.

Watkins, P. (2011). Garment pattern design and comfort. In G. Song (Ed.), *Improving comfort in clothing* (pp. 245–277). Cambridge: Woodhead Publishing Limited.

Watkins, S. (1977). The design of protective equipment for ice hockey. *Home Economics Research Journal*, *5*(3), 154–166.

Watkins, S. (1995). *Clothing: The portable environment* (2nd ed.). Ames, IA: Iowa State University Press.

Whitwell, J. (n.d.). Rowing biomechanics: What constitutes optimal, efficient technique? Retrieved from https://rowingbiomechanics.weebly.com/.

Wibowo, A. W. D., Wijayanto, T., Widyastuti, W., & Herliansyah, M. K. (2018). The Effect of Clothing Fit and Material of Women's Islamic Sportswear on Physiological and Subjective Responses during Exercise in Warm and Humid Environment. *MATEC Web of Conferences, ICET4SD 2017*. 154, 01074.

Wright, G. R., Nicoletti, J., & Shephard, R. J. (1977). Selection, training and development of youth oarsmen. In H. Lavalläe, & R. J. Shephard (Eds.), *Frontiers of activity and child health* (pp. 293–305). Quäbec City: Editions du Pälican.

Yan, Y., Yi, X., Tao, J., Lin, M., & Li, R. (2011). Research on the influence of the fabric organization to wind-resistant property of sportswear. *Advanced Materials Research, 331*, 646–649.

Yoga bra design for elderly women

9

Newman Lau[a], *Jun Zhang*[a], *Joanne Yip*[b], *Winnie Yu*[b]
[a]School of Design, The Hong Kong Polytechnic University, Hung Hom, Hong Kong;
[b]Institute of Textiles and Clothing, The Hong Kong Polytechnic University, Hung Hom, Hong Kong

9.1 Basic requirements for yoga bra design

Yoga combines spiritual, mental, and physical practices that promote a healthy lifestyle including emotional balance, greater awareness of the self, management of type II diabetes mellitus, and all sorts of other physiological and psychological parameters (Cowen & Adams, 2005; Aljasir, Bryson, & Al-shehri, 2010; Rocha et al., 2012). There are two parts to yoga. On the one hand, yoga involves exercises that control breathing called pranayama, which requires practitioners to inhale and exhale to their most significant extent (Morley, 2001). On the other hand, yoga also involves carrying out various body poses called asanas which include bending in different directions, twisting the body parts with each other or stretching and balancing the body parts (Tran, Holly, Lashbrook, & Amsterdam, 2001). In carrying out these breathing exercises and poses, the yoga bra has utmost importance due to the involved physical exertion and stretching. A bra that is able to help with these areas would enhance performance and recovery from the poses (Troynikov et al., 2010).

The main objectives of traditional yoga bras are to help with the performance of the practitioner during yoga and prevent exercise-induced injuries to the breasts. Therefore, the requirements for the yoga bra design usually focus on function, comfort, and support. Most conventional yoga bra brands in the market claim that their products are able to protect the breasts to prevent breast-related injuries and stretchable to allow a large range of body movement. Nowadays, yoga bras offer far more than these basic functions. They also need to be fashionable and feminine as well as cater to different lifestyles (O'Sullivan, Hanlon, Spaaij, & Westerbeek, 2017). However, an inappropriate fit and poor design can contribute to substantially negative health impacts, such as breast and back pain, or even pain of the entire upper body, which largely affect performance and reduce the enthusiasm of women to take part in exercise (Wood, Cameron, & Fitzgerald, 2008). More seriously, women with large breasts might have to undergo reduction mammoplasty after their breast skin is stretched beyond recovery by an ill-fitting bra (Greenbaum, Heslop, Morris, & Dunn, 2003).

The main structural components of a typical sports/yoga bra are shown in Fig. 9.1. The bra usually consists of structural components including the shoulder straps, back strap, strap slider, strap ring, bra cup, wings, side seams, underband, and hooks.

Those components contribute to the performance of the entire bra by shaping the body, providing support, allowing the wearer to feel comfortable, and eliciting

Latest Material and Technological Developments for Activewear. http://dx.doi.org/10.1016/B978-0-12-819492-8.00009-0

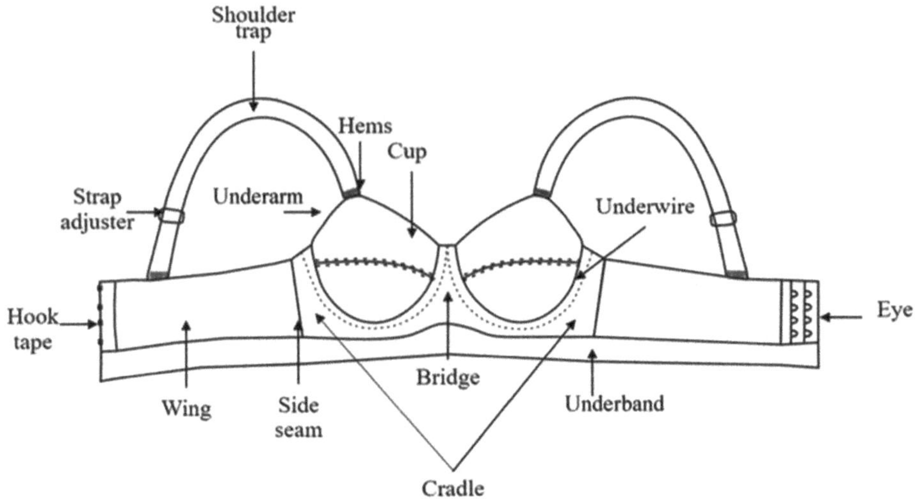

Figure 9.1 Main structural components of a typical sports/yoga bra.

other perceived feelings such as feeling athletic or fashionable. A large number of researches have investigated various kinds of sports bras. The categories of common sports bra are divided into high impact, medium impact, and low impact, which are used for different levels of activities. The low impact sports bras can be regarded as yoga bras which allow more flexibility and body extension. However, there are more information targeted to analyze the performance of yoga bra in terms of its thermal aspects, aesthetics, comfort, support, etc. Besides, some design criteria of sports bra such as thermal comfort, hygienic comfort can also be adopted for yoga bras. Previous research by Gho, Munro, Jones, and Steele (2014) found that the shoulder straps, bra cups, and underwire are the components that commonly cause the most discomfort with sports bras. Among these influential factors, the shoulder straps rank first in discomfort because they are prone to falling off or digging into the shoulders. Discomfort from the bra cups could be because they are gapping, overly loose (wrinkled or puckering cups), or excessively tight (breasts spilling out of bra). Liu et al. (2019) concluded on the major influential factors of comfort of a sports bra. Their work involved 11 different kinds of sports bras and a sample of 31 women with different breast sizes. The results showed that women with varying breast sizes have different concerns on the comfort of a sports bra. To be specific, larger breasted women are more concerned about the shoulder strap design and shaping effect or gathering of the breasts, while women with smaller and medium size breasts mainly focused on the wearability of the bra as well as its ability to shape and lift up the breasts or push them up. Chan, Yu, and Newton (2001) conducted interviews with 80 women to solicit their views on bra fit and support, the cup design, bra closure, sizing system, and elasticity. They found that the underwire, bra cup, and shoulder straps rank as the top three factors of discomfort. The hooks

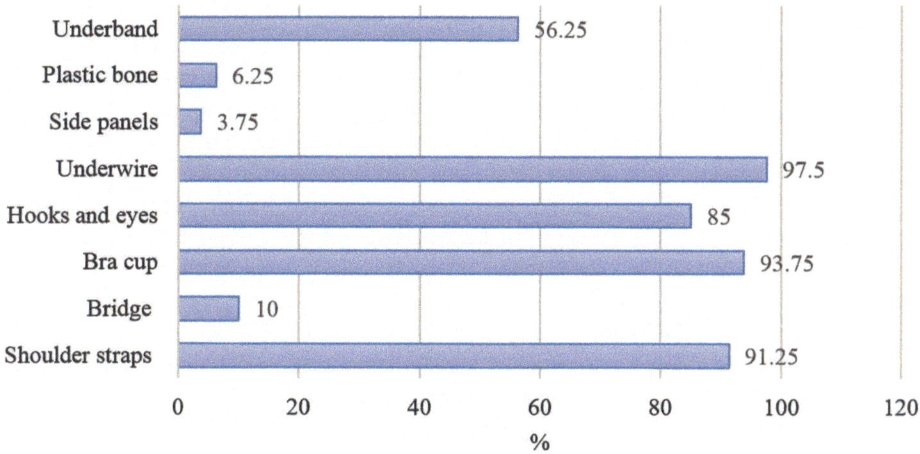

Figure 9.2 Bra components that cause discomfort (Chan et al., 2001).

and eyes as well as the underband design also considerably contribute to discomfort. Fig. 9.2 shows the results of the interviews that identify the bra components that cause the most discomfort.

Zhang, Yu, and Lau (2017) recommended that bra designs for large breasted women need to consider the spilling out of the breasts over the bra cups and gathering of the breasts. Therefore, the materials that are used for the bra cups should have less elasticity or high modulus to provide all round support (Mills, Sanchez, & Scurr, 2016). A bra with a well-structured underband provides the necessary support to the breasts and not the strap orientation (Coltman, McGhee, & Steele, 2015; Steele, 2013). A well-fitting underband is neither too tight nor too loose because the former would compress the chest thus restricting breathing while the latter would not provide adequate support to the breasts. A method to alleviate this problem is to choose fabrics with suitable elasticity. The elasticity of the underband not only accommodates the upper torso movement but also prevents excessive breast motion. The underband should be elastic enough to allow for the chest to expand in the horizontal plane for breathing, but at the same time, minimizes breast motion in the medial-lateral and vertical directions. Unfortunately, all of these required factors make it difficult to modify the design of sports bras. In general, the mechanical properties of the bra fabric not only should allow a lower tensile modulus to prevent large deformation but should also offer good recovery.

There has been an increasing number of research works that focus on designing sports/yoga bras or propose design criteria for them based on advanced and systematic scientific approaches. A significant proportion of the works examine the relationship between the biomechanics of the breasts and perceived bra comfort, support, and aesthetics. Analyses of the biomechanics of the breasts are mainly interested in the kinematic and kinetic data. In the former, much of the evidence points to the relationship between breast displacement during exercise and breast discomfort, pain and other problems. Chen, Gho, Wang, and Steele (2016) investigated the effects of different sports bra and speed of gait based on perceived breast comfort and discomfort by controlling breast motion. They indicated that compression sports bras can provide more support than encapsulation sports bras during running. However, the performance of compression sports bras falls short of encapsulation sports bras when the subjects are stationary.

Moreover, breast acceleration has also been moderated by the support of sports bras, as well as reduces the intensity of the exercise activity (Ainsworth et al., 2000; Greenbaum et al., 2003). McGhee, Power, and Steele (2007) quantified breast motion relative to the trunk by placing infrared light-emitting diodes that are 2 mm in diameter on the nipples of the study participants. The motion capture equipment collected and recorded data on the movement of the diodes and then calculated the relative movements of the nipples. The sports bras selected for the experiment all provided high-levels of support of the breasts. However, there is little difference with breast support for controlling breast displacement in the vertical direction. It was found that the vertical breast displacement appears to be positively related to instantaneous vertical breast velocity (McGhee & Steele, 2010). Originally, they proposed the use of elastic materials for the bra cup to allow a longer downward trajectory of the breasts. However, stride cadence had to be taken into consideration. As a result, there was little effect in using elastic material to limit instantaneous vertical breast velocity. The uplifting of the breasts might be due to donning a sports bra so that there is the combined support of the anatomical support system of the breasts and external support of the sports bra.

With regard to the kinetic measurements of the breasts, Haake, Milligan, and Scurr (2013) converted kinematic data from their female subjects who ran on a treadmill to dynamic strain. After differentiating the dynamic strain twice, they obtained the maximum dynamic acceleration. They concluded that breast discomfort highly depends on the value of dynamic strain and the maximum dynamic acceleration, and reducing breast discomfort during running should aim to decrease their values. In contrast, when a woman is standing upright in a static position, an anterior torque is created on the thorax due to the net force when the woman is bare breasted (Greenbaum et al., 2003; McGhee, Steele, Zealey, & Takacs, 2013). This makes it difficult to determine the nonloaded breast position because the breasts are subjected to gravity. To address this problem, Mills et al. (2016) countered the force of gravity on the breasts by submerging the breasts of their subjects into water and soybean oil, and used the buoyant forces from the fluid to measure the nipple position when gravity is loaded onto the breasts in the static position. They then estimated the neutral nipple position based on the immersion. It was found that gravity plays an important role in affecting the static nipple position in the inferior and posterior directions of the breasts and neglecting the deformation of the breasts by gravity would lead to erroneous results when evaluating the origins of breast pain.

Lawson (1985), Lorentzen and Lawson (1987), and Lawson and Lorentzen (1990) evaluated sports bras on their performance in controlling displacement, and providing overall comfort and support for different breast sizes by using both quantitative and qualitative methods. They concluded that the sports bra that is rated as most comfortable to wear is constructed of knit fabric with a medium modulus, cotton and spandex blend, no fasteners, and a cross-back or Y-back strap configuration, and low modulus fabric inserts around the bra cup that contributed to freedom of movement. Zhou, Yu, and Ng (2013) identified effective design features of sports bras for women between 24 and 40 years old by measuring their breast motion without a bra and with different sports bras donned. They concluded that the most effective features of sports bra designs include a high neckline, cross back, wide straps, etc.

9.2 Design and development of new yoga bra for older women

Yoga wear for elderly women has been drawing more and more attention not only from the academia but also the industry who is interested in commercializing the garment for the sportswear market because there is an increasing number of older women who are more regularly engaging in yoga (Ferrari, 2007; Newson & Kemps, 2007) as they gain awareness of the enormous health benefits from yoga exercise. Although it is now becoming more common knowledge that older women are gravitating toward yoga-related activities, there is little information about yoga bra preferences. Regarding yoga bra designs, there is very few details related to older women. The publications related to the needs of older women in terms of a yoga bra generally focus on three areas: (1) changes in physical characteristics due to ageing, (2) general preferences or demands of older women for yoga bras based on subjective views and physical factors, and (3) responses to the different design criteria of yoga bras.

To obtain a better understanding of the needs of older women and yoga bras, the first and foremost goal is to understand how the ageing process influences them which can then be used to address the latest demands of older women for yoga wear. The ageing process has substantial effects not only on the changes in their body parameters but also on their purchasing behavior. Therefore, older women can be encouraged to purchase and wear a yoga bra, and be more involved in yoga exercises, if the yoga bra design caters to both their specific psychological and physical needs.

9.2.1 Psychological needs

Women in different life stages may be involved in different kinds of exercise, sports and physical activities (O'Sullivan et al., 2017). Previous research has shown that older women who participate in physical exercise do so because they are influenced by health-related reasons, and to enrich their post-retirement life (Cortis, Sawrikar, & Muir, 2007; Leone & Ward, 2013). The factors that significantly motivate older women to do more exercise are confidence, identity, social interaction, external motivations (rewards or goals or family support), and inclusive attire (Caldwell et al., 2018; Kolt, Driver, & Giles, 2004; Qiu & Hu, 2014; Sawrikar & Muir, 2010). Risius, Thelwell, Wagstaff, and Scurr (2014) found that the sports bra is capable of improving the self-confidence of mature women because the bra shapes their body and covers the signs of ageing. More importantly, advanced sports bra designs have combined systematic and technologically advanced means of fiber selection, and structurally supportive with multifunctional designs to create well-designed sports bras that may reduce breast discomfort, prevent exercise-related breast injuries as well as reduce potential permanent damage to breast tissues (Bowles, Steele, & Munro, 2012). Confidence in both appearance and ability to perform yoga would also be increased when older women wear an age-friendly yoga bra.

9.2.2 Physical needs

Eliakim et al. (1996) examined the preferences of older women on sportswear and suggested that the fabric should have pre-shrunk properties and free of strain. Besides, greater elasticity of sportswear especially in the horizontal direction is also important, and a well-designed pattern can appear to reduce the size of their body. Extra pockets are desirable to accommodate personal accessories namely glasses, tissues, etc. The sportswear must be easily donned and doffed because most of the older women have some form of functional disability. Older women prefer front closures in garments to help them don clothing. Zhang, Yu, and Lau (2017) conducted a questionnaire survey and based on the findings, proposed specific design criteria for the bra design of older women. They suggested that older women prefer a high underarm, wider shoulder straps, thin cup padding, and user-friendly open-closure bra for ease of donning and doffing, and more coverage of their body to conceal the back fat and at the same time, control the amount of exposed flesh (breasts).

The ageing process has the most impact on the body proportions, especially in the upper torso of women along with posture changes. Ulger, Erdogan, Kumanlioglu, and Unur (2013) measured the density of the breasts of 120 women and indicated that between post-menopausal and pre-menopausal women, the breast density of the former is drastically reduced. The reduced breast density means a loss of elasticity and therefore that there is less anatomical support of the breasts which causes the breasts to sag under gravity (Ezure, Hosoi, Amano, & Tsuchiya, 2009). Post-menopausal women who are overweight (high BMI) with large breasts more frequently experience thoracic pain. That is, post-menopausal women are affected by a spectrum of anthropometrical changes, which potentially result in biomechanical changes and contribute to physical pain in the thoracic spine (Spencer & Briffa, 2013).

Ashdown and Na (2008) examined the postural changes in the upper torso due to the ageing process by comparing the 3D body scans (as shown in Fig. 9.3) of a group of older women (40 women over 55 years old) and a group of younger women (40 women between 19 and 35 years old). The results clearly showed that significant differences between the breast configurations of the older and younger women. Relatively greater bust point and between-bust point distances can be found among the older women. Moreover, the angle of the upper anterior thoracic of the older women is smaller than that of the younger women. In conclusion, the breasts of the older women sag more and are spaced further apart. These cause more frustrations during bra purchasing which is reflected in their attitude toward buying a bra. To cope with the signs of ageing, older women prefer bras with better uplift ability that gives

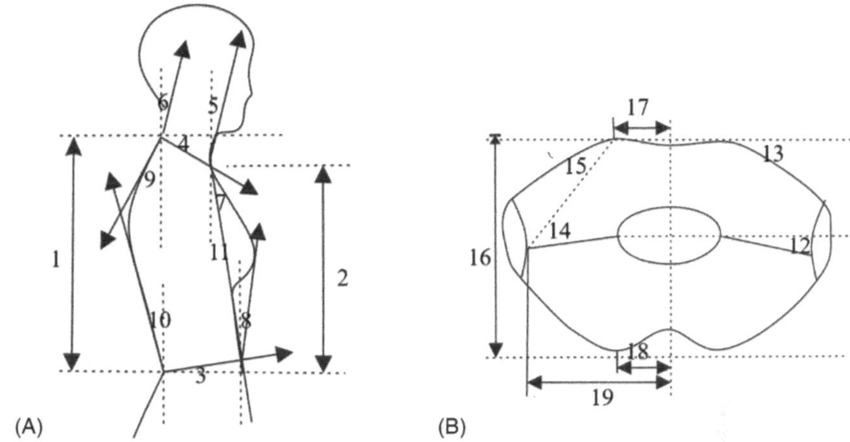

Figure 9.3 Sagittal and transverse views of body measurements based on 3D body scanning. *Notes*: 1 = Length of centre back; 2 = Length of centre front; 3 = Waist angle; 4 = Cervical fossa angle; 5 = Anterior cervical angle; 6 = Posterior cervical angle; 7 = Upper anterior thoracic angle; 8 = Lower anterior torso angle; 9 = Upper posterior thoracic angle; 10 = Lower posterior torso angle; 11 = Upper anterior torso angle. 12 = Shoulder line angle; 13 = Blade angle; 14 = Line of neck point, acromial point and scapular point; 15 = Midscapular point; 16 = Depth of scapular point level;17 = scapular point to centre back; 18 = Bust point to centre front; and 19 = Acromial point to centre front
Source: Ashdown and Na (2008).

more support and subsequently alleviates the problem of breast sagging (Risius et al., 2014; Shin, 2014).

While ageing causes postural changes in the upper torso of older women, ageing also causes changes to the internal composition of their breasts. The breast sagging and wide spacing between the breasts are signs of changes in the internal composition of the breasts. The former is due to excessive stretching of the Cooper's ligaments with time and gravity, which cause the inferior lateral migration of the breast tissues (Patrick, 2005). The average amount of adipose and glandular tissues in the breasts is reduced and can even level off for women between 55 and 69 years old (Gefen & Dilmoney, 2007). As a result, breast motion and forces exerted on the breasts of older women are more significant than those in the case for younger women and therefore their extent needs to be better understood for an ergonomically designed yoga bra.

However, experiments that involve real life subjects have their limitations. Experiments are complicated preparation procedures, require subject recruitment, and are costly and time consuming but produce only a limited amount of data. On the other hand, a large and increasing number of studies use simulation methods such as finite element modeling to address the shortcomings of experiments and surveys. The flowchart for constructing a finite element model is shown in Fig. 9.4. Nonlinear finite element simulation can be used as an approach to analyze breast deformation under both a braless condition and when a bra is worn. There are mainly three threads of research on the modeling of breast deformation with the finite element method.

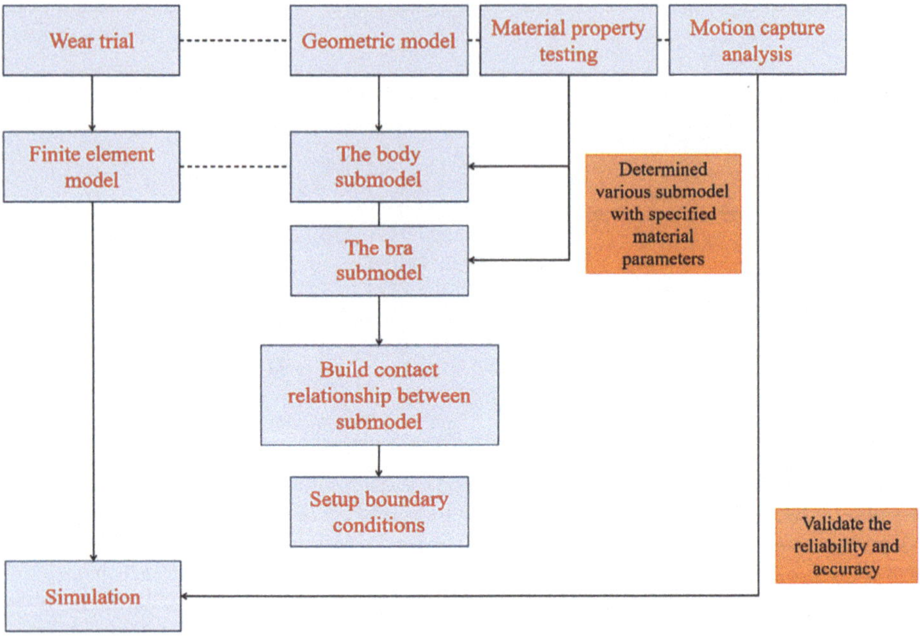

Figure 9.4 Flowchart of constructing finite element model.

Representative examples that use different technologies to build the finite element model are shown in Table 9.1. Aside from the differences in obtaining the geometric model and its applications in the various finite element models, there are also some research studies that focus on the boundary conditions. Gamage, Boyes, Rajagopal, Nielsen, and Nash (2012) created a finite element model and set the boundary conditions to predict breast deformation from a prone to supine position on the upper body under the effect of gravity. However, considering the large amount of deformation on the breasts, they found that it is necessary to partition the model into more detailed and accurate mechanical and anatomical parameters, such as stiffness. Chen, Ng, Yu, Zhou, and Wan (2013) separated the skin into three layers with specific mechanical properties combined with a thorax and two breasts to simulate the large deformation of the breasts by using a hyper-elastic neo-Hookean material model to examine the effects of jogging on the breasts.

The finite element model built to analyse the breast skin extension during yoga pose Warrior II (Fig. 9.5) can be used for ergonomic design of a yoga bra (Zhang, Sun, Lau, Yip, & Yu, 2019). This finite element model was constructed by a simplified musculoskeletal system in upper women body. By using this model, the movements of upper body could be simulated more realistic and accurate.

The upper torso of an older woman was divided into several components in the simulation model including the irregularly shaped pectoralis major muscles, humerus, breasts, skin, and other supporting bones based on the human anatomical structure. Regarding the material properties of the different components in the model, a

Table 9.1 Examples of different methods to build finite element model for biomechanical analysis of breasts

Method			
	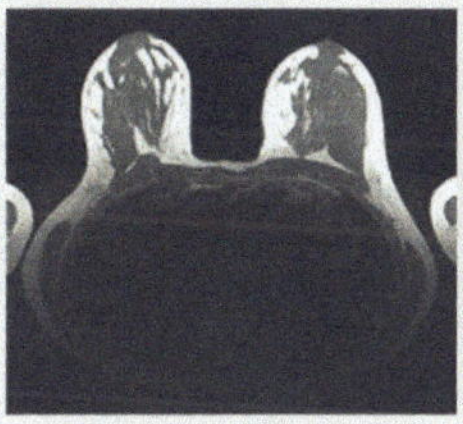	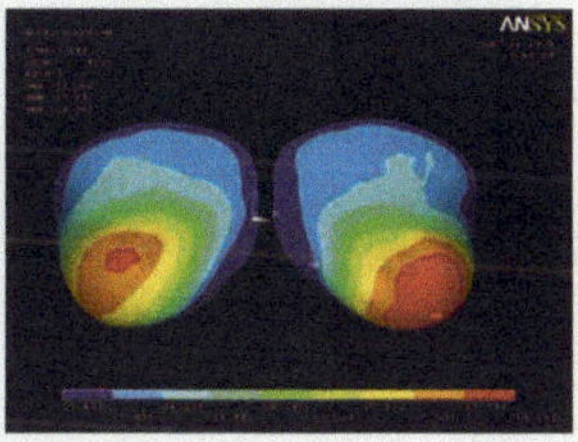	
Geometric model acquisition method	Magnetic resonance imaging (MRI; 3D X-ray microscopy and gamma-ray imaging)	Tailored 3D photogrammetric imaging system for body	3D laser body scanner to capture body images in different bra conditions
Characteristics	Highly accurate; internal tissues and proportions can be viewed in supine and prone positions	Predict breast displacement during highly intensive activities	Predict breast motion with braless and different bra conditions
Application	Diagnostic information for surgery	Referenced for ergonomic design of yoga bra in this study	Referenced for ergonomic design of yoga bra in this study
Source	Danch-Wierzchowska, Borys, and Swierniak (2016)	Chen et al. (2013)	Sun et al. (2019)

Figure 9.5 Yoga pose "Warrior II".

hyper-elastic material model based on the Mooney-Rivlin model was adopted to the breast tissues, and neo-Hookean model was adopted to represent the elasticity of the skin. The simulated results during arm abduction were then compared to the motion capture data based on the breast deformation. By making the adjustment on the material properties of the different components in the model, the best matched simulation

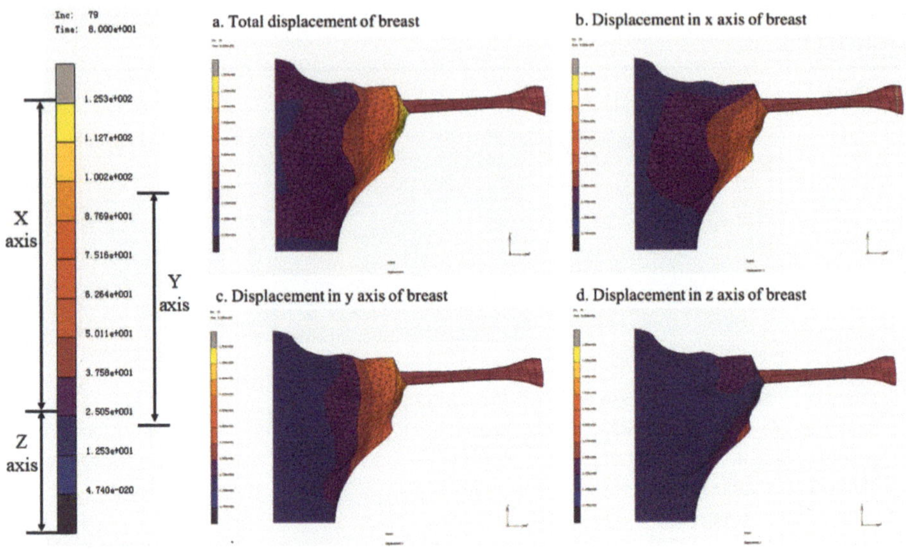

Figure 9.6 Finite element modeling of upper musculoskeletal system to simulate breast deformation during arm abduction.

could be obtained. The material parameters of the pectoralis major muscles were then determined.

The results simulated by the finite element model are presented in Fig. 9.6. It shows that the upper-outer quarter of the breast had the largest extension during arm abduction (Zhang et al., 2019). The extension value is almost four times larger than that of the upper-inner quarter of the breast. This may be attributed to the contraction of the inner pectoralis muscle. As the subjects undergo the aging process, there is loss of elasticity of their muscles. Therefore, the breasts of older women may experience more deformation than those of younger women while doing yoga. It is therefore important to ensure that there is more support in the upper-outer quarter of a yoga bra.

9.2.3 Material selection

The material selection for the yoga bra is an important factor during the design process. The textile materials are chosen based on the intended use and characteristics of the target users. Table 9.2 shows some of the common types of fabrics used for sports bras. Most of these fabrics are synthetics with a high percentage of spandex to provide good stretch and recovery.

To determine which fabrics should be selected and how to configure for various components of yoga bra, some essential properties of the fabrics should be evaluated, including thermal balance, moisture management, durability, tactile sense, and some mechanical properties such as elasticity, bendability, recoverability, and compression ability. Zhang (2004) suggested that it is important to examine the thermal properties of garment materials for clothing of older people such as in sportswear material

Table 9.2 Common types of fabrics used for sports bras in the market

Fabric No.	Fabric sample	Magnification	Fiber content	Textile structure	Properties
1		 	80% Nylon 20% Spandex	Knitted honeycomb	Good stretch ability and recovery, soft, accommodates body shape
2		 	78% Nylon 22% Spandex	1 × 1 rib knit	Durable, good stretch ability and recovery
3		 	85% Polyester 15% Spandex	Single Jersey	Lightweight, quick drying, easy to wash and care
4	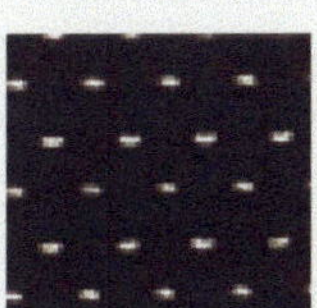	 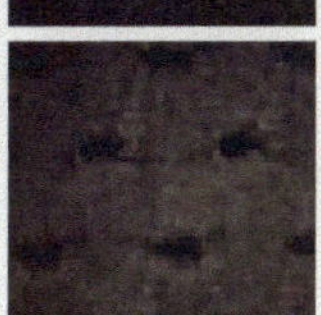	80% Nylon 20% Spandex	Warp knitted eyelet	Quick drying and good air permeability, wicks moisture

(Continued)

Table 9.2 Common types of fabrics used for sports bras in the market *(Cont.)*

Fabric No.	Fabric sample	Magnification	Fiber content	Textile structure	Properties
5		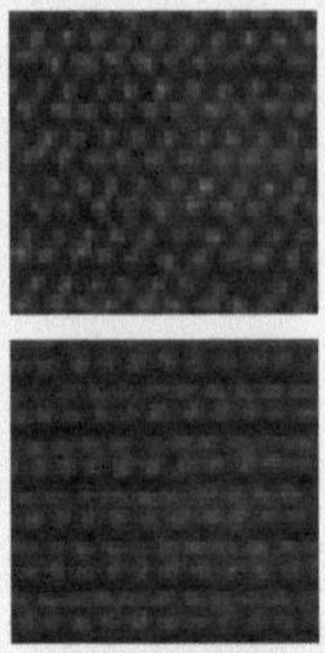	90% Nylon 10% Spandex	Single Jersey	Good stretch ability and recovery, soft, accommodates body shape, lightweight
6		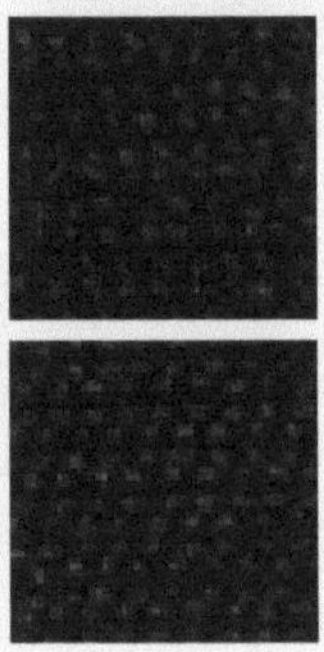	93% Polyester 7% Spandex	1 × 1 rib knit	Easy to wash and care, less elastic

selection because they are sensitive to external thermal changes. The use of specialized thermal textile materials in sportswear for mature individuals, for example, polypropylene, is recommended to prevent their body from hypothermia and avoid skin irritation. Layered clothing also provides better insulation.

The moisture property of textile material is defined by the water absorption rate at the top and bottom layers of the fabric, the speed in which moisture spreads and by using the one-way transport index. Moisture management capability can be graded in an order from 1 to 5, which ranges from poor to excellent (Krenzer, Starr, & Branson, 2005). With regard to choosing the yoga bra material for older women, outstanding moisture management capability of the bra fabric is important, as this kind of fabric allows the wearer to feel dry (and comfortable) by absorbing or wicking the sweat away quickly. Fabric Samples 1 and 4 in Table 9.2 are considered to have good water and air permeabilities due to their porous structure. As a result, these two kinds of fabrics are usually used for the armhole area or anywhere else that would require air movement and vapor permeance; that is, perspiration to be wicked away.

Fabrics used for yoga bras also need to have durability. The yoga bra might continuously rub against the torso and arms during yoga. Accordingly, some of the body parts might be vulnerable to abrasion (Venkatraman, 2015). The amount of abrasion caused by a textile is related to its fiber contents and mechanical properties, quality of the fibers or yarns, designed structure, and density and count of the fibers or yarns.

The yoga bra is also an intimate apparel which is likened to a second skin, so that the hand can be obtained immediately and the wearer can quickly determine whether the material feels good against her skin (Yu, 2011). The fabric hand is usually tested for softness, stiffness, smoothness, itchiness, prickliness, roughness, crispness, etc. (Klatzky & Lederman, 2010). Traditionally evaluated by subjective methods such as a questionnaire, the results were usually inconsistent because perceptions of the same textile fabric often varied among individuals. However, there are also some common suggestions to provide a good handle for bras. Plain single jersey knitted fabric has a smooth surface and 20% drape coefficiency, which means that it is relatively more drapeable and less stiff (drapeable to rigid ranges from a drape coefficiency that is close to 100% to zero) (Venkatraman, 2015). Hence, Fabric Samples 3 and 5 in Table 9.2 are considered to be suitable for use as the outer layer of a yoga bra and a t-shirt bra design. To avoid skin irritation because the breasts of older women are less elastic and dense, components such as lace, accessories and embroidery should be applied by using sew-free technologies. These would include seamless molded cups (Wu, Yick, Ng, & Yip, 2011), thermoplastic adhesive materials for the elastic bra band (Covelli, Liu, & Farmer, 2007) or ultrasonic bonding for hemming purposes (Gaul, Schneider-Ramelow, & Reichl, 2009).

9.2.4 Design sketches

Clothing pressure is one of the most impacting factors on the overall perceived comfort of bras. The pressure cannot be too excessive nor inadequate. Zhang, Yeung, and Li (2002) and Hong, Dongsheng, Qufu, and Ruru (2011) recommended a pressure between 60 and 100 g/cm^2 as a pressure over 100 g/cm^2 would cause discomfort because it would exceed the pressure in the capillary vessels. Some adverse physiological effects would result, such as reduced cardiac output, reduced blood flow, and even cardiac sympathetic activity (Brown et al., 2010; Watanuki & Murata, 1994). However, if the clothing pressure is excessively low, the garment is not able to carry out its intended function nor does it allow for more movement efficiency. Excessively high- or low-pressure material is not desirable for age-friendly yoga bras. Older women require a yoga bra with stronger textile materials for the bra cup to lift up their sagging breasts and create a more aesthetically pleasing illusion. The pressure level of the materials used for different parts of the body such as the abdomen, breast, or chest, should have different elasticity to accommodate the body curvatures (Wu & Li, 2018). The overall pressure comfort of a yoga bra should not depend on one specific area, but the distribution of the pressure over the entire garment (Wang, Luo, Li, Dai, & You, 2005).

It has been well-documented that sports bras can prevent excessive movement of the breasts during physical activities (Zhou, Yu, Ng, & Hale, 2009). During exercise, the bra and breasts interact. To control breast motion, the movement of energy through the breasts should be absorbed by the yoga bra. This shock energy absorbance may be achieved by both the changes in the shape of the various bra components. For example, the bra cup can act as an energy absorber. When there is external loading onto the bra cup, a bra cup with good compression ability can transfer the kinetic potential energy to the cup by changing its shape. The unloading takes place when the cup reverts back to its original shape. The load transfer of the breast motion to the bra then transfers back to the breasts. The breast motions are therefore controlled, and the breasts are protected by the bra. However, each time that there is a transfer of energy, the breast suffers from continuous bouncing (Regina & Foggiatto, 2018). Therefore, older women with large breasts would benefit from the use of wider bra straps coupled with bra cups that have good compressibility.

Some of the design criteria for a yoga bra for elderly women are proposed and shown in Table 9.3 based on not only the results and findings of the aforementioned literature but also the functions of each component of the bra.

Fig. 9.7 shows the design specifications of the yoga bra in this study, including adjustable straps and closure, higher neckline, tailor-made molded cup for the elderly, etc. The prototype was successfully produced; see Table 9.4 for details.

Table 9.3 Design criteria of age-friendly yoga bra

Component	Design criterion	Characteristic
Fabric	Breathable High thermal comfort	■ Natural fabric that provides cooling effect, like jade infused fabric ■ Odor resistance polyester knitted fabric ■ Breathable and thin materials
Elastic bra band strap	No stretching Easy to adjust Easy to wear Easy to move	■ Adjustable straps and closure ■ Front adjusting strap ■ No digging ■ Strong elastic with good recovery
Neckline	No exposure	■ Higher neckline
Back	Allows large movements Conceals back fat Cannot be a cross-back or pull-over bra	■ Back design: Narrow panel
Armhole	Allows large movements	■ Concave panel
Underbust curve	Conceals fat	■ Wider underband
Cup	Round breast shape No sagging	■ Molded tailor-made bra cup for elderly
Side panel	Lifts up the breasts	■ Special cut to lift up and support the breasts
Aesthetics	Fashionable Gives younger appearance	■ Dark or pastel colors ■ Long shirt to cover ageing body ■ Color matching
Closure	Easy to don Easy to doff	■ Adjustable front closure
Wing	More coverage	■ Wider wing design

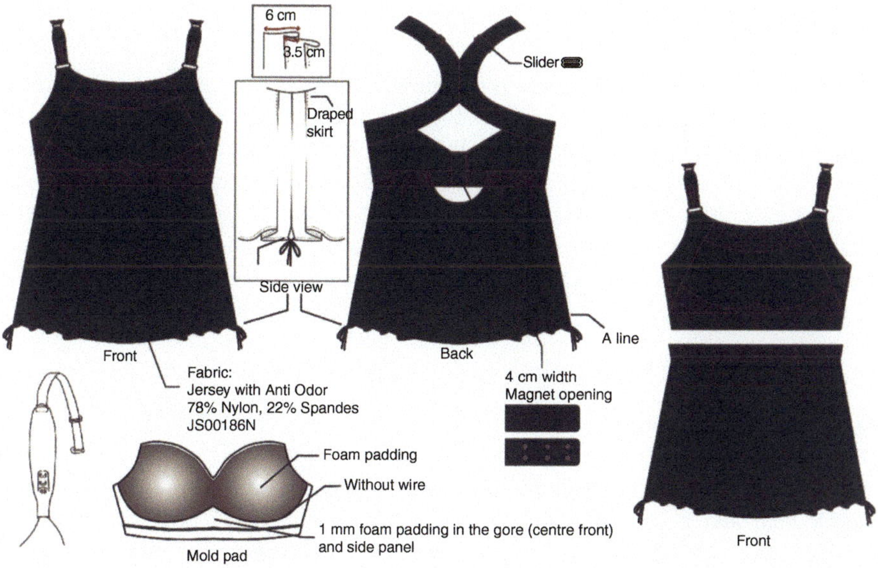

Figure 9.7 Design specifications of yoga bra for elderly women.

Table 9.4 Design prototype and characteristics

- Front-adjusting shoulder straps with slider
- Butterfly-shaped back design allows large movements
- Option to wear as single sports bra or sports vest
- Magnetic openings enable ease of donning and doffing
- A-line design of the vest provides slimming effect, hides the belly and is comfortable to wear

9.3 Conclusion

In conclusion, the yoga bra designed for mature women in this study takes into consideration good uplifting ability to support sagging breasts, and patterns and colors that contribute to a fashionable and aesthetically pleasing design to improve the body shape illusion. The bra also has high side seams, good elasticity, and good shaping ability, allows ease of donning and doffing, and has excellent air permeability and moisture wicking properties, seam-free bonded design, and smooth fabric surface. Large breasted mature women will benefit from the encapsulation style of the bra, wider and stronger shoulder straps, and bra cups with the appropriate amount of compressibility.

9.4 Future plans

Although the yoga bra designed and developed in this study can address the primary psychological and psychological needs of older women, yoga bras in general can offer more than that. A good yoga bra encourages exercise so that older women are more inclined to be physically active and therefore stay healthier for a longer time. Although individual preferences and needs may dictate the type of yoga bra used, the ultimate goal is to gain a better understanding and explain to the older female population on ways to be more physically active, including social influences and how they affect physical activity attitude. These in turn influence yoga bra purchase and wear behavior. Socialization comes from parents, family, teachers and the community in general. Therefore, perspectives and views can be changed through lived experiences. This is especially true among one's peers which is the basis for mutual influence. As a result, the attitude of older women toward exercise and yoga bra use can be influenced by the perceptions, values, personality, and attitude of other people.

It is recommended that sports bras be designed as a form of intervention, and the design needs to focus on the in-group identity of older women to influence their attitude around exercise and specifically yoga. In doing so, this may change their internal cognitive process around identity formation; specifically on the value of exercise and prioritization of health and well-being by wearing a yoga bra. That is, there will be two identities: women who wear a yoga bra during yoga and are considered to be healthy and physically active and women who do not wear one, and are not considered to value exercise as much as their peers. This could also be grounded into an appropriate theory to motivate older women to exercise by using an approach that incorporates cultural factors into an interactive relationship between individuals and their social surroundings.

References

Ainsworth, B. E., Haskell, W. L., Whitt, M. C., Irwin, M. L., Swartz, A. M., Strath, S. J., ..., & Jacobs, D. R. (2000). Compendium of physical activities: An update of activity codes and MET intensities. *Medicine and Science in Sports and Exercise*, *32*(9), S498–S504.

Aljasir, B., Bryson, M., & Al-shehri, B. (2010). Yoga practice for the management of type II diabetes mellitus in adults: A systematic review. *Evidence-Based Complementary and Alternative Medicine*, *7*(4), 399–408.

Ashdown, S. P., & Na, H. (2008). Comparison of 3-D body scan data to quantify upper-body postural variation in older and younger women. *Clothing and Textiles Research Journal*, *26*(4), 292–307.

Bowles, K. A., Steele, J. R., & Munro, B. J. (2012). Features of sports bras that deter their use by Australian women. *Journal of Science and Medicine in Sport*, *15*(3), 195–200.

Brown, A. D., McMorris, C. A., Longman, R. S., Leigh, R., Hill, M. D., Friedenreich, C. M., & Poulin, M. J. (2010). Effects of cardiorespiratory fitness and cerebral blood flow on cognitive outcomes in older women. *Neurobiology of Aging*, *31*(12), 2047–2057.

Caldwell, A. E., Masters, K. S., Peters, J. C., Bryan, A. D., Grigsby, J., Hooker, S. A., ..., & Hill, J. O. (2018). Harnessing centred identity transformation to reduce executive function burden for maintenance of health behaviour change: The Maintain IT model. *Health Psychology Review*, *12*(3), 231–253.

Chan, C. Y., Yu, W., & Newton, E. (2001). Evaluation and analysis of bra design. *The Design Journal*, *4*(3), 33–40.

Chen, L. H., Ng, S. P., Yu, W., Zhou, J., & Wan, K. F. (2013). A study of breast motion using non-linear dynamic FE analysis. *Ergonomics*, *56*(5), 868–878.

Chen, X., Gho, S. A., Wang, J., & Steele, J. R. (2016). Effect of sports bra type and gait speed on breast discomfort, bra discomfort and perceived breast movement in Chinese women. *Ergonomics*, *59*(1), 130–142.

Coltman, C. E., McGhee, D. E., & Steele, J. R. (2015). Bra strap orientations and designs to minimise bra strap discomfort and pressure during sport and exercise in women with large breasts. *Sports Medicine Open*, *1*(1), 21.

Cortis, N., Sawrikar, P., & Muir, K. (2007). *Participation in sport and recreation by culturally and linguistically diverse women*. Sydney, Australia: Social Policy Research Centre, University of New South Wales.

Covelli, C., Liu, H., & Farmer, D. (2007). *U.S. Patent Application No. 11/745,668*.

Cowen, V. S., & Adams, T. B. (2005). Physical and perceptual benefits of yoga asana practice: Results of a pilot study. *Journal of Bodywork and Movement Therapies*, *9*(3), 211–219.

Danch-Wierzchowska, M., Borys, D., & Swierniak, A. (2016). Breast deformation modeling based on MRI images, preliminary results. In *Proceedings of Internation Conference of Information Technologies in Biomedicine* (pp. 227-234). Springer, Cham.

Eliakim, A., Brasel, J. A., Mohan, S., Barstow, T. J., Berman, N., & Cooper, D. M. (1996). Physical fitness, endurance training, and the growth hormone-insulin-like growth factor I system in adolescent females. *Journal of Clinical Endocrinology & Metabolism*, *81*(11), 3986–3992.

Ezure, T., Hosoi, J., Amano, S., & Tsuchiya, T. (2009). Sagging of the cheek is related to skin elasticity, fat mass and mimetic muscle function. *Skin Research and Technology*, *15*(3), 299–305.

Ferrari, C. K. (2007). Functional foods and physical activities in health promotion of aging people. *Maturitas*, *58*(4), 327–339.

Gamage, T. P. B., Boyes, R., Rajagopal, V., Nielsen, P. M. F., & Nash, M. P. (2012). Modelling prone to supine breast deformation under gravity loading using heterogeneous finite element models. In P. Nielsen, A. Wittek, & K. Miller (Eds.), *Computational biomechanics for medicine* (pp. 29–38). New York, NY: Springer.

Gaul, H., Schneider-Ramelow, M., & Reichl, H. (2009). Analysis of the friction processes in ultrasonic wedge/wedge-bonding. *Microsystem Technologies*, *15*(5), 771–775.

Gefen, A., & Dilmoney, B. (2007). Mechanics of the normal woman's breast. *Technology and Health Care*, *15*(4), 259–271.

Gho, S. A., Munro, B. J., Jones, S. C., & Steele, J. R. (2014). Evidence-based recommendations for building better bras for women treated for breast cancer. *Ergonomics*, *57*(5), 774–786.

Greenbaum, A. R., Heslop, T., Morris, J., & Dunn, K. W. (2003). An investigation of the suitability of bra fit in women referred for reduction mammaplasty. *British Journal of Plastic Surgery*, *56*(3), 230–236.

Haake, S., Milligan, A., & Scurr, J. (2013). Can measures of strain and acceleration be used to predict breast discomfort during running? In *Proceedings of the Institution of Mechanical Engineers. Part P: Journal of Sports Engineering and Technology*, *227*(3), 209–216.

Hong, L., Dongsheng, C., Qufu, W., & Ruru, P. (2011). A study of the relationship between clothing pressure and garment bust strain, and Young's modulus of fabric, based on a finite element model. *Textile Research Journal*, *81*(13), 1307–1319.

Klatzky, R. L., & Lederman, S. J. (2010). Multisensory texture perception. In *Multisensory object perception in the primate brain* (pp. 211–230). New York, NY: Springer.

Kolt, G. S., Driver, R. P., & Giles, L. C. (2004). Why older Australians participate in exercise and sport. *Journal of Aging and Physical Activity*, *12*(2), 185–198.

Krenzer, G., Starr, C., & Branson, D. (2005). Excellence in design-development of a sports bra prototype. *Clothing & Textiles Research Journal*, *23*(2), 131–134.

Lawson, L. (1985). A Comparison of Eight Selected Sports Bras: Biomechanical Support, Overall Comfort Ratings and Overall Support Ratings. Retrieved from https://digitalcommons.usu.edu/etd/4039/.

Lawson, L., & Lorentzen, D. (1990). Selected sports bras: Comparisons of comfort and support. *Clothing and Textiles Research Journal*, *8*(4), 55–60.

Leone, L. A., & Ward, D. S. (2013). A mixed methods comparison of perceived benefits and barriers to exercise between obese and nonobese women. *Journal of Physical Activity and Health*, *10*(4), 461–469.

Liu, K., Zhang, L., Zhu, C., Zhao, X., Lu, W., Li, M., & Wang, J. (2019). An analysis of influence factors of sports bra comfort evaluation based on different sizes. *The Journal of The Textile Institute*, *110*(12), 1792–1799.

Lorentzen, D., & Lawson, L. (1987). Selected sports bras: A biomechanical analysis of breast motion while jogging. *Physician and Sportsmedicine*, *15*(5), 128–139.

McGhee, D. E., Power, B. M., & Steele, J. R. (2007). Does deep water running reduce exercise-induced breast discomfort? *British Journal of Sports Medicine*, *41*(12), 879–883.

McGhee, D. E., & Steele, J. R. (2010). Breast elevation and compression decrease exercise-induced breast discomfort. *Medicine and Science in Sports and Exercise*, *42*(7), 1333–1338.

McGhee, D. E., Steele, J. R., Zealey, W. J., & Takacs, G. J. (2013). Bra-breast forces generated in women with large breasts while standing and during treadmill running: Implications for sports bra design. *Applied Ergonomics*, *44*(1), 112–118.

Mills, C., Sanchez, A., & Scurr, J. (2016). Estimating the gravity induced three dimensional deformation of the breast. *Journal of Biomechanics*, *49*(16), 4134–4137.

Morley, J. (2001). Inspiration and expiration: Yoga practice through Merleau-Ponty's phenomenology of the body. *Philosophy East and West*, 73–82.

Newson, R. S., & Kemps, E. B. (2007). Factors that promote and prevent exercise engagement in older adults. *Journal of Aging and Health*, *19*(3), 470–481.

O'Sullivan, G. A., Hanlon, C., Spaaij, R., & Westerbeek, H. (2017). Women's activewear trends and drivers: A systematic review. *Journal of Fashion Marketing and Management: An International Journal*, *21*(1), 2–15.

Patrick, C. W. (2005). Engineering adipose tissue for regenerative and reparative therapies. *Seminars in Plastic Surgery*, *19*(3), 207–215.

Qiu, C., & Hu, Y. (2014). The academic thinking based on the perspective of fashion design. *International Journal for Innovation Education and Research*, *2*(1), 17–26.

Regina, L., & Foggiatto, J. A. (2018). Bra fitting - a Brazilian mastectomized women experience. In *Proceedings of International Conference on Applied Human Factors and Ergonomics* (pp. 213-222). Springer, Cham.

Risius, D., Thelwell, R., Wagstaff, C. R. D., & Scurr, J. (2014). The influence of ageing on bra preferences and self-perception of breasts among mature women. *European Journal of Ageing*, *11*(3), 233–240.

Rocha, K. K. F., Ribeiro, A. M., Rocha, K. C. F., Sousa, M. B. C., Albuquerque, F. S., Ribeiro, S., & Silva, R. H. (2012). Improvement in physiological and psychological parameters after 6 months of yoga practice. *Consciousness and Cognition*, *21*(2), 843–850.

Sawrikar, P., & Muir, K. (2010). The myth of a 'fair go': Barriers to sport and recreational participation among Indian and other ethnic minority women in Australia. *Sport Management Review*, *13*(4), 355–367.

Shin, K. (2014). Intimate apparel: Designing intimate apparel to fit different body shapes. In *Designing apparel for consumers* (pp. 273–291). Cambridge, UK: Woodhead Publishing.

Spencer, L., & Briffa, K. (2013). Breast size, thoracic kyphosis & thoracic spine pain-association & relevance of bra fitting in post-menopausal women: A correlational study. *Chiropractic & Manual Therapies*, *21*(1), 20.

Steele, J. R. (2013). The biomechanics of better bras: improving support and comfort during exercise. In *ISBS-Conference Proceedings Archive* (Vol. 1, No. 1).

Sun, Y., Yick, K. L., Yu, W., Chen, L., Lau, N., Jiao, W., & Zhang, S. (2019). 3D bra and human interactive modeling using finite element method for bra design. *Computer-Aided Design*, *114*, 13–27.

Tran, M. D., Holly, R. G., Lashbrook, J., & Amsterdam, E. A. (2001). Effects of Hatha yoga practice on the health-related aspects of physical fitness. *Preventive Cardiology*, *4*(4), 165–170.

Troynikov, O., Ashayeri, E., Burton, M., Subic, A., Alam, F., & Marteau, S. (2010). Factors influencing the effectiveness of compression garments used in sports. *Procedia Engineering*, *2*(2), 2823–2829.

Ulger, H., Erdogan, N., Kumanlioglu, S., & Unur, E. (2003). Effect of age, breast size, menopausal and hormonal status on mammographic skin thickness. *Skin Research and Technology*, *9*(3), 284–289.

Venkatraman, P. (2015). Evaluating the performance of fabrics for sportswear. In *Materials and technology for sportswear and performance apparel* (pp. 272–299). Boca Raton, FL: CRC Press.

Wang, J. M., Luo, X. N., Li, Y. I., Dai, X. Q., & You, F. (2005). The application of the volumetric subdivision scheme in the simulation of elastic human body deformation and garment pressure. *Textile Research Journal*, *75*(8), 591–597.

Watanuki, S., & Murata, H. (1994). Effects of wearing compression stockings on cardiovascular responses. *The Annals of Physiological Anthropology*, *13*(3), 121–127.

Wood, K., Cameron, M., & Fitzgerald, K. (2008). Breast size, bra fit and thoracic pain in young women: A correlational study. *Chiropractic & Osteopathy*, *16*(1), 1.

Wu, L., Yick, K. L., Ng, S. P., & Yip, J. (2011). Shape characterization for optimisation of bra cup moulding. *Journal of Fiber Bioengineering and Informatics*, *4*(3), 235–243.

Wu, P. Y., & Li, Y. M. (2018). Research and emphasis on young women's demand for sports bra. *Journal of Business Administration Research*, *7*(1.), .

Yu, W. (2011). Achieving comfort in intimate apparel. In *Improving comfort in clothing* (pp. 427–448). Cambridge, UK: Woodhead Publishing.

Zhang, H. (2004). *Activewear for Older Women who Exercise Regularly: A Product Development Approach.* Retrieved from http://mspace.lib.umanitoba.ca/handle/1993/17945.

Zhang, J., Sun, Y., Lau, N., Yip, J., & Yu, W. (2019). FE model for biomechanical analysis and prediction of older women breast deformation. *International Journal of Management and Applied Science*.

Zhang, S., Yu, W., & Lau, N. (2017). Design criteria for age-friendly bras. In *Proceedings of 14th Asian Textile Conference* (pp. 170-173).

Zhang, X., Yeung, K. W., & Li, Y. (2002). Numerical simulation of 3D dynamic garment pressure. *Textile Research Journal*, *72*(3), 245–252.

Zhou, J., Yu, W., & Ng, S. P. (2013). Identifying effective design features of commercial sports bras. *Textile Research Journal*, *83*(14), 1500–1513.

Zhou, J., Yu, W., Ng, S. P., & Hale, J. (2009). Evaluation of shock absorbing performance of sports bras. *Journal of Fiber Bioengineering and Informatics*, *2*(2), 108–113.

An innovative tank top equipped with biofeedback system for adolescents with early scoliosis

10

Joanne Yip, Garcia Kwok
Institute of Textiles and Clothing, The Hong Kong Polytechnic University, Hung Hom, Hong Kong

10.1 Introduction

The human body is made up of bones, which provide the supportive framework for the body. In this framework, the spine is an essential component that supports the human body. It is located at the center of the back of the body, which significantly contributes to maintaining posture during normal activities and also serves as the central nerve transmission channel. Unfortunately, some individuals, especially adolescent girls, may suffer from a spinal deformity called scoliosis that has no known cause. The condition affects their daily life and in extreme cases, even the quality of their life suffers (Morgan & Scott, 1956). The Oxford Concise Color Medical Dictionary (Martin, 1998) defines scoliosis as the "lateral deviation of the backbone". Severe scoliosis ultimately affects the social life of patients. They are unable to perform normal activities, like walking, speaking or taking part in sports due to their spinal deformity. Most cases of scoliosis are idiopathic (Hewitt, 2011), which means that the cause of the condition is unknown. The most current type of therapy for scoliosis is one that is passive. Patients with less severe scoliosis, which means that the spine has a 10–20° curvature, undergo regular check-ups every 6–12 months, and their doctor may suggest exercise or physiotherapy as part of the treatment process. For scoliotic patients with a spinal curvature that exceeds 21–45°, bracing therapy is usually prescribed (Prosthetic And Orthotic Unit, 2005) with the use of a brace made of rigid plastic material.

The effectiveness of a spinal orthosis, whether it is a hard or flexible brace, depends on its ability to remind the wearer about her/his spinal curvature through increased pressure exerted onto certain points of the body. This would *serve to alert the child of poor posture as well as motivate him/her to straighten her/his spine*. Therefore, different related types of treatments, such as physical therapy, chiropractic care, biofeedback and electric stimulation, are now being used to manage scoliosis.

In particular, biofeedback is an area of growing interest in medicine and psychology because it has proven effective to address a number of physical, psychological and psychophysical problems (Ahmed, Begum, Funk, Xiong, & von Scheele, 2011; Lehrer, Smetankin, & Potapova, 2000; Nestoriuc, Martin, Rief, & Andrasik, 2008). Biofeedback therapy is a nonmedical process that involves the measuring of specific and quantifiable bodily functions of a subject, such as the brain wave activity, blood

Latest Material and Technological Developments for Activewear. http://dx.doi.org/10.1016/B978-0-12-819492-8.00010-7

pressure, heart rate, skin temperature, sweat gland activity, and muscle tension, thus conveying the information to the patient in real-time. The basic aim of biofeedback therapy is to support a patient in realizing her/his own ability to control specific psychophysiological processes (Kappes, 2008). Wong, Mak, Luk, Evans, and Brown (2001) studied the effectiveness of feedback devices in postural training for adolescent idiopathic scoliosis (AIS) patients. The results showed a curve control rate of 69%.

Wong and Wong (2008) launched a smart garment that can monitor the posture of the trunk. The garment integrates accelerometers and gyroscopes which can detect postural changes in terms of the changes in the curvature of the spine in the sagittal and coronal planes. However, the specific and quantifiable bodily functions of a subject, such as muscle tension, are not taken into consideration in this system. Also, the data acquisition and feedback system are not wireless and the battery used to power the device is relatively bulky. There is therefore substantial room for improvement if this sort of audio-biofeedback system is used for posture training.

The literature has consistently concluded that surface electromyography (sEMG) biofeedback is effective for muscle rehabilitation. A review of the sEMG studies on upper extremity dysfunctions of the physically disabled (Lyons et al., 2003) has concluded that sEMG is an important means of increasing the muscle activity of the upper extremities and most effective when used in conjunction with physiotherapy. In a meta-analysis of studies that applied sEMG biofeedback to hemiplegic stroke subjects, Schleenbaker and Mainous (Schleenbaker & Mainous, 1993) concluded that sEMG improves functional outcomes in both the upper and lower extremities, and therefore should be incorporated into therapeutic regimes.

In this chapter, the aim is to discuss the design and development of a posture training tank-top for 10–13 year old pre-teen and teenage girls who have mild scoliosis, in order to train them into obtaining a relatively more balanced muscle activity profile and posture. The posture training tank-top is fabricated with textile materials and incorporates 3-axis accelerometers that provide real-time monitoring to alert the wearer about her posture.

10.2 Literature review

The current literature on scoliosis is subsequently reviewed to provide some basic knowledge on this condition. The symptoms and means of identifying scoliosis are elaborated to provide insight into the diagnosis of this condition and its recommended treatments. These include conservative bracing with a flexible or rigid brace, as well as the related problems including the effectiveness and other concerns, which are discussed to illuminate the potential issues and possible means of improvement. In addition, other related methods of posture correction are discussed to provide an understanding on the importance of maintaining a "good" posture and the impact of posture on scoliosis patients. Besides, a biofeedback system, which is the main monitoring system in this project, is studied in terms of its general applications, and applications related to posture training. Then, their advantages will be incorporated into a tank top that encourages posture training. A posture monitor sensor will also be developed that works well on the designed tank-top.

10.2.1 Introduction on scoliosis

"Scoliosis" is a term that originated from the ancient Greek word "skoli-osis", which means crooked. Scoliosis is a condition in which the spine has a lateral curvature in the anteroposterior plane (Webster, 1983). More recently, researchers have concluded that cases in which the lateral curvature of the spine measures 10° or more based on radiographs can be diagnosed as scoliosis (Cassella & Hall, 1991; Kane, 1997). In general, scoliosis can be classified into three main types, namely, idiopathic, congenital or secondary to a neuromuscular disease. Idiopathic scoliosis is a structural spinal curvature that occurs without any apparent reason (Reamy & Slakey, 2001). Congenital scoliosis takes place before or at birth. Scoliosis cases that are secondary to a neuromuscular disease take place when the spinal curvature is caused by the impairment of the muscles or the direct nervous system control is affected. Agabegi and Agabegi (Agabegi & Agabegi, 2005) indicated that the approximate distribution of the three types of scoliosis is 65% idiopathic, 15% congenital and 10% secondary to a neuromuscular disease. Hewitt (Hewitt, 2011) provided a similar distribution, with 65% idiopathic, 15% congenital and 20% other reasons. Therefore, the focus of this work will be on idiopathic scoliosis as this condition affects the mass population of patients.

Idiopathic scoliosis can be further classified by the age of the patients at initial diagnosis. Juvenile idiopathic scoliosis patients are first diagnosed between the ages of 4 and 10 years old (Coillard, Circo, & Rivard, 2010). AIS patients are first diagnosed between the ages of 10 and 15 years old and have skeletal maturity (Dobbs & Weinstein, 1999). AIS is the most common type of scoliosis. There has been a number of research work done that indicates rapid physical growth during puberty is one of the factors that contributes to the progression of the spinal curvature (Lonstein & Carlson, 1984; Mackenzie, 1922). Besides, scoliosis tends to progress more rapidly in girls than boys (Roach, 1999). Pehrsson et al. (Pehrsson, Larsson, Oden, & Nachemson, 1992) indicated that the girl to boy ratio for scoliosis is 1:1 between 4 and 6 years old, 2:1 to 4:1 between 6 and 10 years old, and 8:1 for those over 10 years old. This also indicates that girls require treatment more than boys based on the affected population.

10.2.2 Scoliosis treatment

The treatment options suggested to scoliosis patients are based on their spinal situation. In general, periodic observations every 6 or 12 months are suggested to scoliosis patients with a spinal curvature of 10–20°. It is recommended that patients with a spinal curvature of 21–45° wear a non-invasive brace that would confine the body posture and prevent the progression of the spinal curvature. For patients with a spinal curvature over 45°, surgery is recommended which involves the fusion of the spinal bones or vertebrae together. The fusion is stimulated by bone grafting of the vertebrae, that is, filling the vertebrae with bone that is grafted from the ribs or pelvis. After fusion, the spine is unable to bend thus straightening the spine (Dolan, Donnelly, Spratt, & Weinstein, 2007; USC Center for Spinal Surgery, 2005). However, Morgan and Scott (Morgan & Scott, 1956) pointed out that this type of surgery will restrict the skeletal maturity and growth of AIS patients, and in extreme cases, skeletal maturity and growth are inhibited as the fused area of the spine may not grow together.

Conventional braces, also known as rigid braces, are prescribed for scoliosis patients who have a spinal curvature of 21°–45°. Conventional braces are tailor-made to accommodate the shape of the body of the wearer by using rigid plastic materials. There are internal brace pads added to specific points to create points of pressure. The objective of conventional braces is to stabilize the spine by exerting pressure onto the affected regions of the thoracic spine and the trunk. The trunk movement will also be restricted. The scoliosis patient then has to wear the brace for 23 h a day for 3–4 years until skeletal maturity (Dworkin et al., 1985). The thoracolumbar sacral orthosis (TLSO) brace (Fig. 10.1), which uses a three-point pressure system to correct spinal curvature, is one of the most common rigid braces on the market.

Flexible braces correct poor posture by using bands of elastic material. The most popular flexible brace is the dynamic SpineCor brace (Fig. 10.2) which was first produced in 1993. All of the components of the SpineCor brace are less than 1.5 mm in

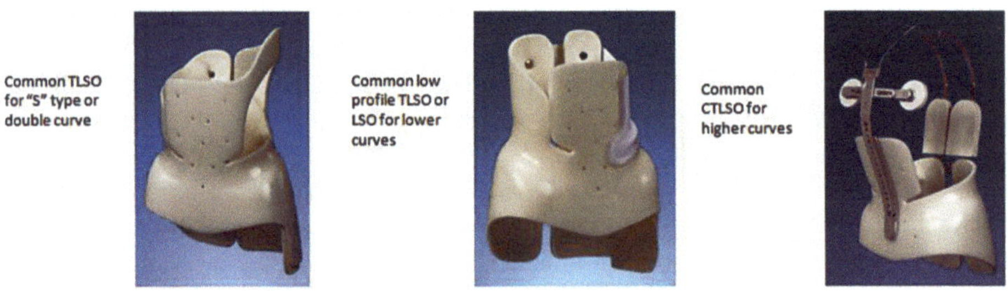

Figure 10.1 Thoracolumbar sacral orthosis (TLSO) brace .

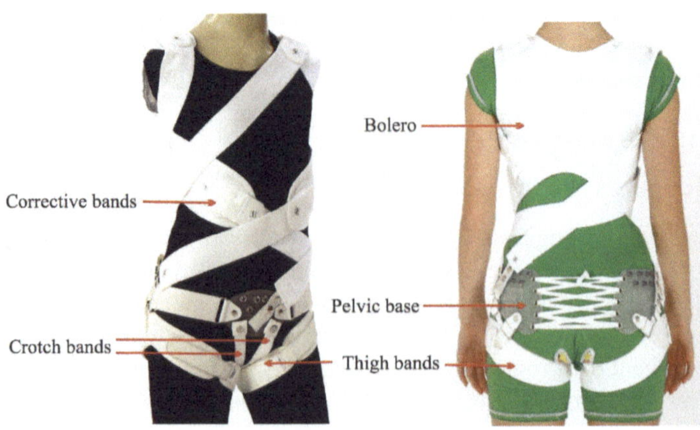

Figure 10.2 SpineCor brace (adapted from www.spinecor.com).

thickness (Weiss & Weiss, 2005). The brace consists of a pelvic base, which is a belt with three pieces of soft thermo-deformable plastic, stabilized by two bands on the thighs and two bands at the crotch, a bolero made of cotton and four corrective bands of elastic material which vary between 0.20 and 1 cm (Wong et al., 2008). The therapy involves mechanical biofeedback or corrective movement (Berger, Edelsten, Fishman, Krebs, & Springer, 1983), in which dynamic harnesses are applied to the patient that control the scoliotic deformity by dynamically controlling the shoulders, thorax, and pelvic girdle, thus limiting adverse movements, and modifying the postural geometry while preserving body movement and growth (i.e., the Corrective Movement© principle) (Berger et al., 1983). The flexible brace allows some controlled movement and the bands of elastic material are adjustable to accommodate the body contours.

10.2.3 Bracing therapy problems

Because of the limitations in trunk motion instigated by wearing a rigid brace, it is possible that there will be atrophy of the spinal musculature and the spine will also become less flexible in the long-term (Bazzarelli et al., 2001). Besides, rigid braces are made of nonbreathable plastic material, and therefore ventilation is greatly reduced which causes lower tolerance of the patient, especially during hot and humid weather (Berger et al., 1983). Rigid braces also limit activities and attract unwanted attention, all of which are significant issues for adolescents (Bunge, de Bekker-grob, van Biezen, Essink-bot, & de Konig, 2009). It is possible that the issues created by wearing a rigid brace have led to poor patient compliance which ultimately causes failure of the treatment (Weiss & Weiss, 2005).

The results of a study conducted by Wong et al. (Berger et al., 1983) showed that there are no differences between the use of a flexible brace and a rigid brace in terms of skin irritation, pressure sores, breathing, and sports participation, but gives pervasive effects on walking, sleeping, and putting on clothing. Some of the wearers of flexible braces find that it is problematic to use the bathroom. For instance, the dynamic SpineCor brace has a plastic shell at the pelvic level of the wearer. This may cause difficulties when patients attempt to bend down.

The most important considerations in brace treatment as described by Bunge et al. (2009) are that scoliosis patients and their families should have confidence in the effectiveness of the therapy, and understand the potential of reducing the risk of surgery and discomfort while wearing the brace. At the same time, brace efficacy, clinical effectiveness, and compliance of the patient are associated as the key factors for the successful treatment of idiopathic scoliosis (Lou, Raso, Hill, Mahood, & Moreau, 2004).

10.2.4 Patient compliance

Aesthetics is one of the most important factors that determine patient compliance (Wickers, Bunch, & Barnett, 1977) but the bulkiness of the brace will affect physical appearance, which is extremely important to adolescents (Dworkin et al., 1985; Fallstrom, Nachemson, & Cochran, 1984; Myers, Friedman, & Weiner, 1970; Wickers et al., 1977). Therefore, bracing may have psychological impacts on patients.

Most adolescent patients are concerned about psychosocial issues, such as peer pressure at school, social relations and lack of family support rather than suffering from scoliosis. These are mostly triggered by individual psychological, negative predictive determinants such as lack of energy, low expectations, anxiety about failure, low self-esteem, and poor body image. Subsequently, compliance with brace wear is affected (Korovessis, Zacharatos, Koureas, & Megas, 2007; Lindeman & Behm, 1999; Morton, Riddle, Buchanan, Katz, & Birch, 2008; Tones, Moss, & Polly, 2006). Therefore, there is a very high possibility that the failure of brace treatment may be due to the lack of patient compliance (Wong et al., 2001). The psychological factors and aesthetics concerns of brace therapy could be more decisive factors than the type of brace (Hasler, Wietlisbach, & Büchler, 2010).

10.2.5 Behavioral methods for compliance enhancement

The educating of patients and their parents on the background information of scoliosis, its natural history and the anticipated risk of curve progression is important (Hasler et al., 2010). Treatment can never be successful without the involvement of the patient (Wong et al., 2001). On top of that, questions on wear time should be asked at each visit and the importance of compliance should be emphasized (Hasler et al., 2010), as the patients and their parents should be educated about all the consequences and possibilities brought on by scoliosis. Aside from education, the level of compliance can be reliably quantified by using temperature loggers to determine the wear time of the brace (Nicholson, Ferguson-pell, Smith, Edgar, & Morley, 2002; Nicholson, Ferguson-pell, Smith, Edgar, & Morley, 2003). This is an essential tool to enhance the treatment efficacy of scoliosis, as subjective assessments with the use of questionnaires, diaries and interviews reveal that there is often the overestimation of time spent in the brace (Morton et al., 2008; Nicholson et al., 2002) compared to the objective data acquired from temperature loggers. The amount of time that the brace is worn and daily wear pattern can be noted based on the data collected by the temperature loggers. Therefore, by using the percentage of hours worn in relation to the prescribed 23-h regimen, the level of compliance is quantified and calculated (DiRaimondo & Green, 1988); over 90% denotes high compliance; 50%–90% denotes moderate compliance, and less than 50% denotes low compliance.

10.2.6 Noninvasive treatment: bracing therapy

Bracing therapy is a noninvasive type of treatment which involves the use of rigid and flexible braces, as well as exercise. With regard to the acceptance level of patients between the rigid and dynamic SpineCor braces, Wong at al. (Wong et al., 2008) conducted a questionnaire and found that the group of subjects who wore the dynamic SpineCor brace encounter more problems when using the bathroom, and subjects who wore the rigid brace face greater difficulties in donning and doffing of the brace. Ólafasson, Saraste, and Ahlgren (1999) compiled statistics that indicated nearly two-thirds of the patients encounter similar difficulties with a rigid orthosis. The brace causes irritation to the wearer, thus resulting in deterioration in the quality of life

Table 10.1 Comparison of treatment effectiveness: rigid brace versus flexible brace

Treatment	Treatment problems
Rigid brace	• Atrophy of spinal muscle (Berger et al., 1983) • Spine becomes less flexible (Berger et al., 1983) • Poor ventilation of brace (Wong et al., 2008) • Psychosocial issues due to brace appearance (Bazzarelli et al., 2001) • Poor compliance (Dworkin et al., 1985)
Flexible brace	• Effectiveness questionable (Weiss & Weiss, 2005) • Causes difficulties when using bathroom (Wong et al., 2008)

(Deceuninck & Bernard, 2012). Coillard et al. (Coillard et al., 2010) performed a test to determine the effectiveness of the SpineCor brace, and the results showed that the brace is effective in treating juvenile idiopathic scoliosis with the ability to correct, stabilize and maintain the Cobb's angle for the long term. However, the spinal curve of most of the adolescent patients who wore the SpineCor increased at least 5° (Weiss & Weiss, 2005). Wong et al. (2008) indicated that both the SpineCor and rigid braces can provide a significant reduction in the Cobb's angle as the immediate effect and after the first 3 months of intervention; however, curve progression still took place for those who wore the SpineCor after 6 months of therapy.

Another important factor to consider when choosing the ideal brace is its effectiveness. In the same study by Wong et al. (2008), the survival rate or clinical efficacy of those who wore the SpineCor brace and the rigid brace is 68% and 95%, respectively. The subjects who wore a rigid brace were more able to finish the therapy compared to the SpineCor brace users.

There are no studies that compare the effectiveness of treatment with a brace versus treatment by exercise. However, the efficacy of treatment by exercise can be summarized by referring to the research scale and the methodology applied. Unlike bracing treatment, the experiment scale of treatment by exercise is relatively small. The number of subjects involved is usually less than 10 (Lau, 2011; Mehta, 1978; Monroe, 2012). Moreover, the method or the form of exercise used as treatment is inconsistent. The problems that concern treatment with rigid and flexible braces, and treatment by exercise are listed in Table 10.1. The problems show that the various types of non-invasive treatments for scoliosis patient need to be improved.

10.3 Importance of correcting posture

10.3.1 Prevalence of poor posture and possible reasons

Juskeliene et al. (Juskeliene, Magnus, Bakketeig, Dailidiene, & Jurkuvenas, 1996) conducted an experiment with 791 children and measured the distance between the seventh cervical vertebra to the lower angles of the left and right scapula. In total, 46.9% of the children were found to have trunk asymmetry caused by poor posture. Oates, Evans, and Hedge (1998) observed the sitting position of 95 children who range from 8 to

12 years old. They found that none of the observed postures are acceptable. Marschall, Harrington, and Steele (1995) conducted a research study in the United States, in which 22.8% of the elementary school students reported backaches, and this percentage increased to 33.3% among the secondary school population. This is resultant of the poorly design school furniture which causes the formation of poor posture habits. Hong, Fong, and Li (2011) studied school bag designs and load on spinal posture, and found that the spine is significantly tilted when children carry a single-strap athletic bag while climbing stairs. When the load of the bag, which can have either double straps or a single strap, exceeds 15% of the body weight of the wearer, the spinal tilt is also significant.

10.3.2 Corrected standing posture

As suggested by McKenzie (1980), the lordosis is reduced in the lower back by standing tall, with the aid of the chest, stomach and buttock muscles. This means that it is not only the back of the body which supports standing, but also the muscles at the back and front of the body, which also help to support the entire body (left one as shown in Fig. 10.3). Zacharkow (1988) supported this finding, and indicated the importance of maintaining a correct standing posture because otherwise, breathing will be affected with reduced lung expansion and the diaphragm is depressed which causes breathing difficulties.

10.3.3 Correct sitting posture

McKenzie (Zacharkow, 1988) also recommended ways to maintain a correct sitting posture. The head and ankles should be straight, shoulders and hips are level, kneecaps face the front and the chin should be parallel to the floor and aligned with the ears. The lower back should be slightly bent forward to support the body with no extra weight distributed onto the spine (corrected one as shown in the right side of Fig. 10.4). The key factor of sitting correctly is to form a lordosis in the lower back. By doing so, the muscles of the lower back can help to support the body while sitting and it is very likely that the spine and the head will be straight.

10.4 Biofeedback systems

10.4.1 Introduction on biofeedback systems

Biofeedback therapy is "a method of training which enables a person mostly with the help of electronic equipment, to learn to control otherwise involuntary bodily functions—'learning to play the internal organs'" (Bray, 1998). This definition means that any instrument can be a biofeedback device as long as it allows the user to learn how to control involuntary body functions. The first recorded use of a biofeedback system was in 1958 by Maia Lisina, who used an electronic device that trained subjects to control their blood vessel contraction and expansion (Gatchel & Price, 1979). Biofeedback has proven to be effective for addressing a number of physical, psychological and psychophysical problems (Ahmed et al., 2011; Lehrer et al., 2000;

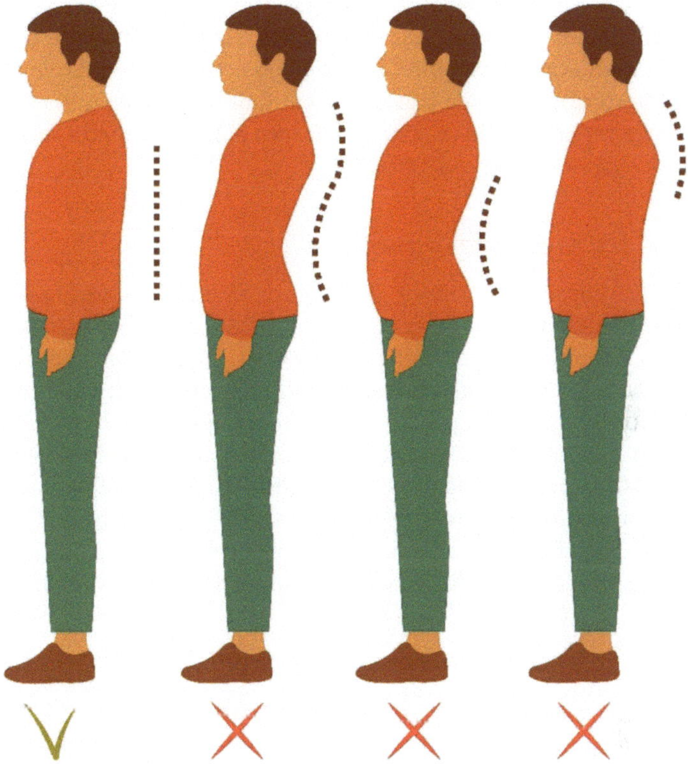

Figure 10.3 Correct standing posture.

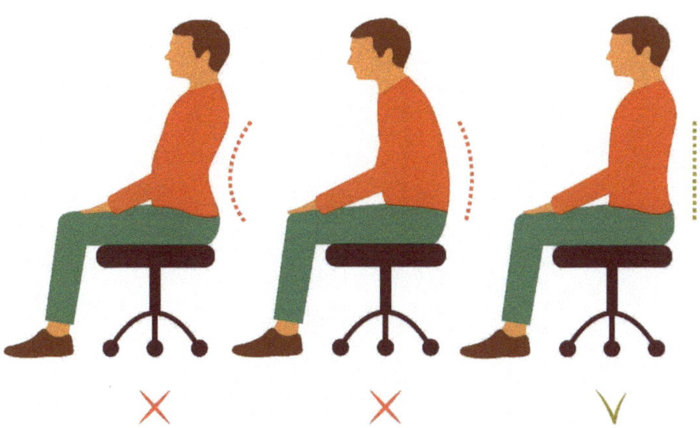

Figure 10.4 Correct sitting posture.

Nestoriuc et al., 2008). It is a nonmedical process that involves the measuring of specific and quantifiable bodily functions of a subject, such as brainwave activity, blood pressure, heart rate, skin temperature, sweat gland activity, and muscle tension, thus conveying the information to the patient in real-time. The basic aim of biofeedback therapy is to support patients in realizing their own ability to control specific psychophysiological processes (Kappes, 2008).

10.4.2 Surface electromyography

The literature has consistently shown that the use of sEMG biofeedback is effective for muscle rehabilitation. A review of studies on sEMG that examine upper extremity dysfunctions in the physically disabled (Lyons et al., 2003) concludes that sEMG is a valuable method for increasing upper extremity muscle activity and most effective when used in conjunction with physiotherapy. In a meta-analysis of sEMG biofeedback studies on hemiplegic stroke subjects, Schleenbaker and Mainous (Matthews & Crawford, 2006) concluded that the use of sEMG improves functional outcomes in both the upper and lower extremities, and that sEMG should be included in therapeutic regimes. Therefore, a thorough evaluation of the myoelectric activity in the muscles of those with AIS is extremely important for sEMG biofeedback training.

10.4.3 Application of biofeedback system on scoliosis subjects

Wong et al. (2001) conducted a study on the effectiveness of audio-biofeedback on the postural training of AIS patients with the use of a device called Micro Straight. The AIS patients were selectively chosen as participants because Micro Straight was a relatively new device at the time. This device uses two loops (Fig. 10.5) made with a nylon fishing line to monitor the length of the torso circumference and respiration from the chest circumference. When the posture of the user is poor, the circumference of the torso length will change and differ from the correct posture that had been set as the standard before the test was carried out. If the user maintains a poor posture for

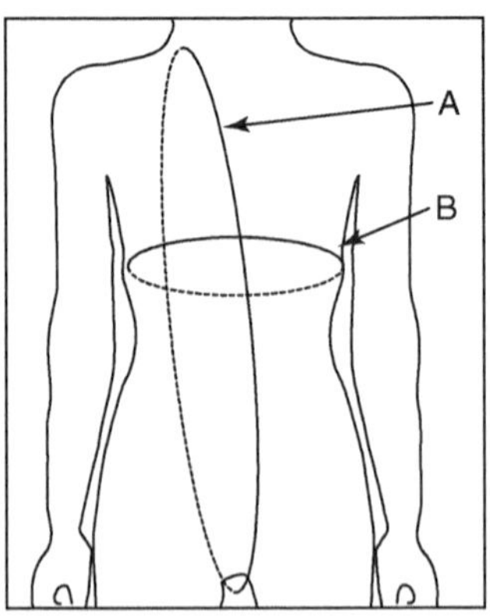

Figure 10.5 Audio-biofeedback in postural training - Micro Straight (Wong et al. 2001).

more than 20 s, the device will emit a sound to alert the user to correct the posture. The volume becomes progressively louder if the user continues to ignore the alert. In addition, the data stored in the device are logged for a period of 4 months to ensure that the progression of the spinal curvature is under control.

The results of the study by Wong et al. (2001) with Micro Straight indicated that the spinal curvature in 9 out of 16 patients is increased less than a Cobb's angle of 5°. The rate of the curvature control is 69% which is similar to the use of the Milwaukee brace to control loading on the concave side of the spine. More importantly, the patients are able to maintain a good posture even after the training regime ended. Wong et al. (2001) also provided suggestions for the further development of posture training for AIS patients. The patients were interviewed to share their experiences on use of the device, and they found that even though they felt fatigued during the first stage of use, the fatigue is reduced as long as they continue to maintain the correct posture. Other than that, the users were satisfied with the use of the device compared to a corrective brace. Besides, the researchers provided suggestions on the audio alert of the device, as the users had to use the device for 23 h each day. The audible tone was overly loud or soft for the users, depending on the situation. Therefore, Wong et al. (2001) suggested the use of vibration over sound even though this would require more battery power.

Wong (2009) also developed a postural training biofeedback device for correcting posture. The results demonstrated the effectiveness of the device. The subjects underwent wear trials of 4 days in length, and during the 4 days, the total time that the subjects showed a poor posture was reduced by about 26%. Although the device is effective in helping patients correct their posture, there are areas that still need to be improved. For instance, the device as shown in Fig. 10.6 requires a relatively large battery for monitoring purposes, and the garment that holds the device requires extra assistance so that there is alignment with the targeted area of monitoring. Besides, the experiment only involved two AIS subjects with one single curve and one double

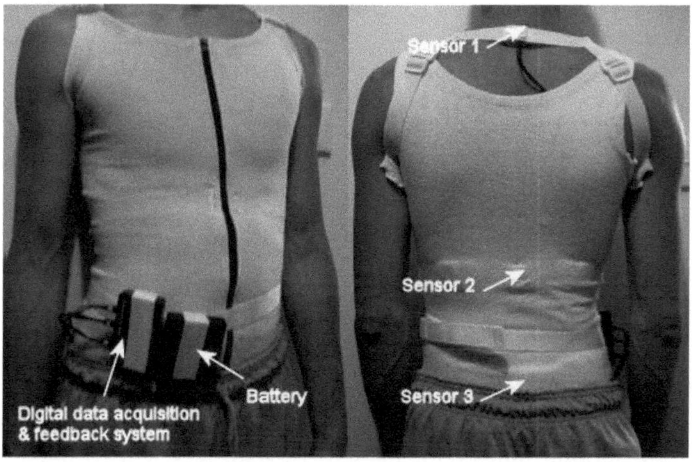

Figure 10.6 Large battery used in postural training biofeedback device for monitoring purposes (Wong 2009).

Table 10.2 Comparison of existing biofeedback posture training devices

Study	Pros	Cons
Effectiveness of audio-bio feedback in postural training for AIS patients (Wong et al., 2001)	• Similar control rate of spinal curve as Milwaukee brace. • Subjects maintain good posture after training ended. • Subjects satisfied with device compared to brace treatment.	• Subjects felt fatigue during the first stage of use. • Audio alert tone overly loud or soft.
Postural training biofeedback device (Wong, 2009)	• Comprehensive data logging system on posture. • Effective for correcting posture of subjects.	• Large and immobile battery size. • Extra assistance required for positioning device. • Experiment was only 4 days and longer duration is recommended.

curve, respectively. Therefore, the effectiveness of the system on treating spinal deformities has yet to be validated. Nevertheless, the research shows the effectiveness and possibility of posture training with EMG monitoring.

Different studies have also been carried out on the paraspinal muscle activity of adolescents with idiopathic scoliosis by using sEMG. Avikainen, Rezasoltani, and Kauhanen (1999) plotted the force versus time and sEMG versus time for the paraspinal muscles during repetitions of maximal isometric trunk extensions. The isometric force versus time curves as well as the maximal integrated EMG activity that were recorded from both sides of the thoracic and lumbar spines of normal and scoliotic participants do not show any significant differences.

Chwała, Koziana, Kasperczyk, Walaszek, and Płaszewski (2014) conducted an sEMG assessment of the paraspinal muscles during static exercise with those who have AIS. They found that during symmetric and asymmetric exercises, the muscle tension patterns significantly differ between the normal and scoliotic groups in comparison with the examination at rest, which in most cases, produces positive corrective patterns. A summary of the existing biofeedback posture training devices used in each study is provided in Table 10.2. It is important to study the advantages and drawbacks found in each study since clinical trials on such technology are lacking.

10.5 Experimental design and results

10.5.1 Section summary

In this section, the methods used for developing and evaluating a tank-top equipped with posture monitoring sensors for individuals with AIS will be described. These include the design and development of a posture correction tank-top, fabrication of the tank-top, use of posture monitoring sensors, wear trial and evaluation tests.

10.5.2 Experimental design

The experiment was divided into two phases, namely the design and development of the posture training tank-top for AIS patients, and then the wear trial and evaluation of the posture training tank-top.

The first phase of the experiment involved the design of a tank-top that emphasizes on comfort. Other factors for the tank-top were also taken into consideration, such as durability, and the technical requirements for accommodating posture monitoring sensors on the tank-top. The materials and silhouette of the tank-top were chosen based on their ability to meet the technical requirements.

The second phase of the experiment was a wear trial and evaluation of the tank-top involved twelve subjects who attended about 30 sessions of posture training. The training included 50 min of biofeedback training, and the tank-top with posture monitoring sensors assisted with the training regime. The data collected were statistically analyzed and the significance of the experimental design will be discussed in the following.

10.5.3 Design sketch and description of body mapping for tank-top

The design of the tank top is shown in Fig. 10.7. The tank top should serve two main purposes. First, the tank top should accommodate the posture monitoring sensors and secure their positioning and second, it needs to be a very comfortable garment to wear. In order to realize these two purposes, body mapping is carried out. U-shaped shoulder straps that have a width of 1 inch are used for the back of the tank top, which is a continuous piece that extends from the front to the back of the garment. The shoulder straps provide the functions of (1) exerting a pulling force to secure the tank top in place and (2) maximizing the cooling effect by reducing the fabric coverage at the back. The tank-top also has a panel of powernet fabric on the posterior and medial sides of the torso along the spine and around the chest area (Smith & Havenith, 2011; Smith & Havenith, 2012). The excellent air permeability of the powernet fabric allows these areas which are prone to high rates of perspiration to stay relatively dry and therefore cool. The tank top has a lining in the chest area in which a pair of thin foam paddings can be inserted for increased support of the breasts. The elasticity of the fabric for the foam cup lining is adequate enough to accommodate foam of different thicknesses and size. This design feature allows an adolescent girl to wear the tank-top as an undergarment and no further intimate apparel or bra is required. Therefore, the positioning of the sensors is more secured and close to the skin. The bottom of the lining is attached with a band of elastic for more support.

10.5.4 Material selection and physical testing results

Warp knitted fabrics were chosen over woven and weft knitted fabrics as the shell fabric and lining of the tank-top due to their advantages of increased comfort and dimensional stability. Two types of fabrics and four different fabrics were pre-selected to undergo different physical tests for the lining and shell respectively, because the material composition and structure of the fabrics affect the quality of the resultant

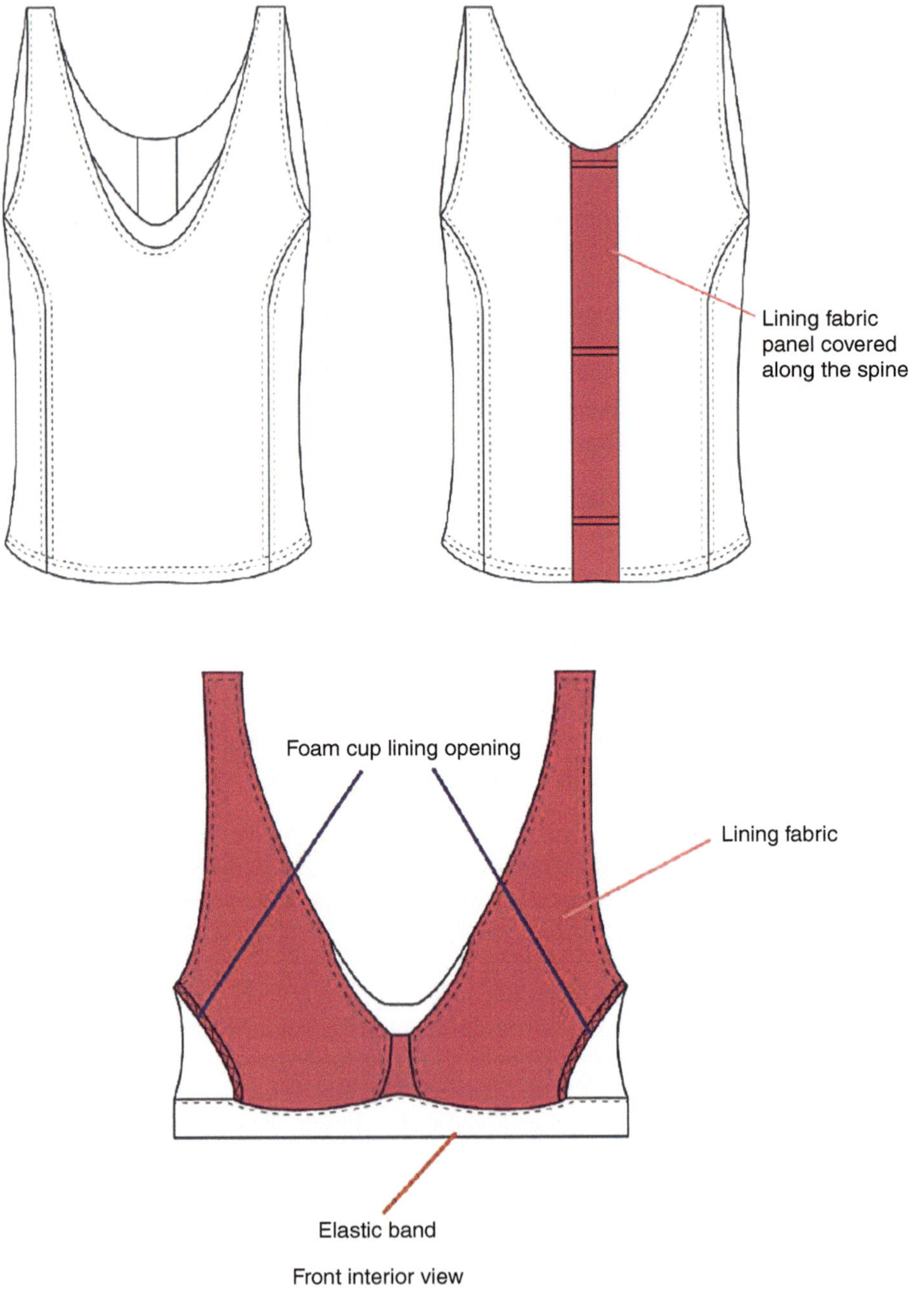

Figure 10.7 Design sketch of body mapping tank top.

garment. Comfort is an important criterion when selecting the fabrics for the tank-top. According to Collier and Epps (1999), the significant factors that affect physical comfort are the ability of the fabric to transfer heat, moisture and air. These may be affected by the structure and thickness of the fabric as these two factors may inhibit the passage of heat, moisture or air. In this tank top, nylon spandex is chosen for the shell and lining because of its exceptional properties, including the ability to dry quickly, lightness in weight, and excellent abrasion and pilling resistance. Besides, the spandex fibers provide extra elasticity for comfort. Table 10.3 shows the fiber composition and specification of different fabrics.

Table 10.3 Materials specifications

Component	**Name**	**Composition**	**Fabric type**	**Density (Wale stitches × Course stitches)**	**Weight (gsm)**	**Thickness (mm)**	**Photo**
Lining Fabric	LI	95% Nylon 5% Spandex	Powemet	44 × 34	80	0.06	
	L2	95% Nylon 5% Spandex	Satinnet	47 × 38	200	0.36	
Shell Fabric	SI	65% Nylon 35% Spandex	Tricot	94 × 70	180	0.28	
	S2	80% Nylon 20% Spandex	Tricot	72 × 70	210	0.62	
	S3	65% Nylon 35% Spandex	Simplex	67 × 70	180	0.32	
	S4	90% Nylon 10% Spandex	Tricot	74 × 90	130	0.2	

Table 10.4 Results of bursting strength testing

	Load (N)				
Fabric name	**Specimen 1**	**Specimen 2**	**Specimen 3**	**Mean**	**SD**
LI	128.1	141.5	168.2	145.9	14.45
L2	150.3	149.7	149.1	149.7	0.429
SI	218.0	220.0	207.5	215.2	4.738
S2	311.3	296.4	307.5	305.1	5.479
S3	362.0	337.6	334.9	344.8	10.57
S4	231.0	236.5	222.8	230.1	4.874

Burst strength testing was performed to determine the mechanical strength of the fabric in accordance with ASTM: D3787-07 (2011) Standard Test Method for Bursting Strength of Textiles- Constant-Rate-of-Traverse (CRT) Ball Burst Test. The test results are shown in Table 10.4.

The load represents the strength required to burst the fabric. A bar chart was subsequently plotted from the results for comparison purposes; see Fig. 10.8. The bar chart shows that the lining fabrics, L1 and L2, have a similar strength. Both require on average about 150 N of force before they burst. Among the three tested specimens acquired from different parts of the fabric, L2 shows a more consistent fabric strength with a lower standard deviation of 0.429, while L1 shows less consistency and a much higher standard deviation of 14.45.

In terms of the samples for the shell fabric, it was found that S1 and S4 have a lower strength than S2 and S3. The load required to burst S1 and S4 is 215.2 N and 230.1 N, respectively. The load for bursting S2 and S3 is over 300 N. All of the shell fabrics demonstrate an average ability to withstand the bursting load with a standard

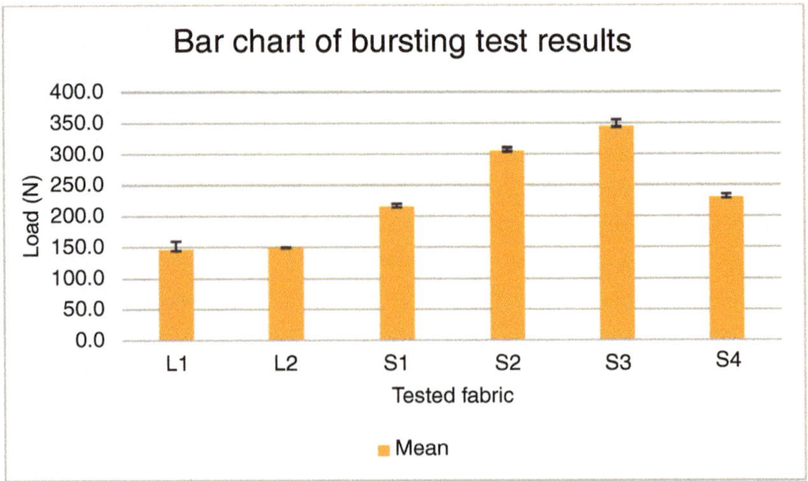

Figure 10.8 Results of bursting strength test of different fabrics.

deviation of about 5. In the bursting test, L2 and S3 have better durability based on their bursting strength than the other tested fabrics.

The stretch and recovery test determines the elongation, recovery ability and elasticity of fabric. The test was conducted in accordance with ASTM: D5035-11 Standard Test Method for Breaking Force and Elongation of Textile Fabric (Strip Method). The specimens were stretched to a tensile strain of 50% and then returned to their original shape. The data were recorded at 10%, 20% and up to 50% in terms of the Newton unit. A lower load required to extend the fabric means higher elasticity of the fabric. The results are shown in Table 10.5.

For ease of comparison, a line chart is provided in Fig. 10.9. The figure shows that L1 has higher elasticity and requires the smallest load to stretch to a tensile strain of 50% followed by L2, S2, S1, S4, and S3 in the warp direction. This is because L1 has 10% more spandex than L2. L2 requires a very high load to deform to a tensile strain of 50% in the weft direction,. This indicates that L2 is a one-way stretch fabric in the warp direction and does not provide a similar amount of stretchability in the weft direction. Therefore, L2 may not be suitable for use as the foam cup lining, since it may not be possible to insert foam cups of a larger size when this fabric is used.

In terms of the shell fabric, the warp and weft directions of the specimens showed similar results. All of the tested fabrics are four-way stretch fabrics which are ideal for use as the shell due to the need for a close fit on the body of the wearer. The test results showed that S3, which has a fiber content of 80% nylon and 20% spandex, has greater elasticity than the other shell fabrics which have more spandex. This is probably due to its stitch density which is slightly lower than that of the other shell fabrics, and thus allows the loops to elongate.

The air permeability test measures the pressure required to push air through tested fabrics. The air permeability of a fabric is higher when a lower pressure is required. The results indicate the breathability of a fabric. When a fabric is very air permeable, the wearer feels cooler as air can flow in and out of the fabric more easily. The air permeability test was conducted in accordance with ASTM: D737-04(2012) Standard Test Method for Air Permeability of Textile Fabric. The results are shown in Table 10.6. A bar chart of the results is plotted for visualization of the results, see Fig. 10.10.

It was expected that a low amount of pressure would be required to push air through the lining fabrics because the mesh properties of powernet and satinnet have rectangular holes between the loops. Therefore, air should easily pass through these holes.

However, L1 has better air permeability than L2 because it has a slightly lower stitch density than L2 and the structure of L1 which is powernet means that the holes are a larger rectangular shape in comparison to L2 which is satinnet. Both of the lining fabrics have better air permeability than the shell fabrics. This shows that powernet and satinnet fabrics are ideal for use as the foam cup lining. The breathability of the tank-top would not be very much affected even if an extra layer of fabric was added to the lining.

In terms of the shell fabrics, S2 has a better performance in comparison to the other shell fabrics. Less pressure (60%) is required to push the air through S2. Therefore, in

Table 10.5 Breaking force and elongation of different textile fabrics

Fabric Name/ Tensile Strain (%)	Elongation Testing Results in Warp Direction Mean Load (N) (S.D.)				
	10%	**20%**	**30%**	**40%**	**50%**
LI	0.26 (0.000)	0.40 (0.000)	0.57 (0.069)	0.71 (0.063)	0.84 (0.075)
L2	0.32 (0.067)	0.59 (0.075)	0.80 (0.000)	1.02 (0.081)	1.20 (0.000)
SI	0.42 (0.014)	0.75 (0.069)	1.07 (0.135)	1.33 (0.136)	1.61 (0.135)
S2	0.37 (0.056)	0.61 (0.036)	0.83 (0.051)	1.07 (0.000)	1.38 (0.075)
S3	0.59 (0.040)	1.08 (0.125)	1.56 (0.081)	2.09 (0.070)	2.64 (0.075)
S4	0.44 (0.064)	0.75 (0.075)	1.07 (0.135)	1.34 (0.145)	1.78 (0.075)
Fabric Name/ Tensile Strain (%)	**Elongation testing results in weft direction Mean load (N) (S.D.)**				
	10%	**20%**	**30%**	**40%**	**50%**
LI	0.21 (0.086)	0.36 (0.088)	0.57 (0.087)	0.76 (0.075)	0.98 (0.081)
L2	1.44 (0.474)	3.47 (0.741)	5.93 (0.805)	8.19 (0.796)	10.6 (0.435)
SI	0.26 (0.000)	0.44 (0.075)	0.63 (0.089)	0.80 (0.130)	1.11 (0.208)
S2	0.48 (0.070)	0.85 (0.090)	1.24 (0.206)	1.69 (0.389)	2.28 (0.462)
S3	0.36 (0.066)	0.67 (0.000)	1.07 (0.000)	1.38 (0.075)	1.92 (0.156)
S4	0.44 (0.075)	0.80 (0.108)	1.17 (0.151)	1.66 (0.193)	2.59 (0.203)

terms of air permeability, S2 is a better option and provides better breathability and allows air to easily pass through. The wearers would be more comfortable if they wear a tank-top fabricated with S2 because they would feel less warm.

The pilling test determines the durability of a fabric. The test was performed in accordance with ASTM D3512/D3512M- 10 Standard Test Method for Pilling

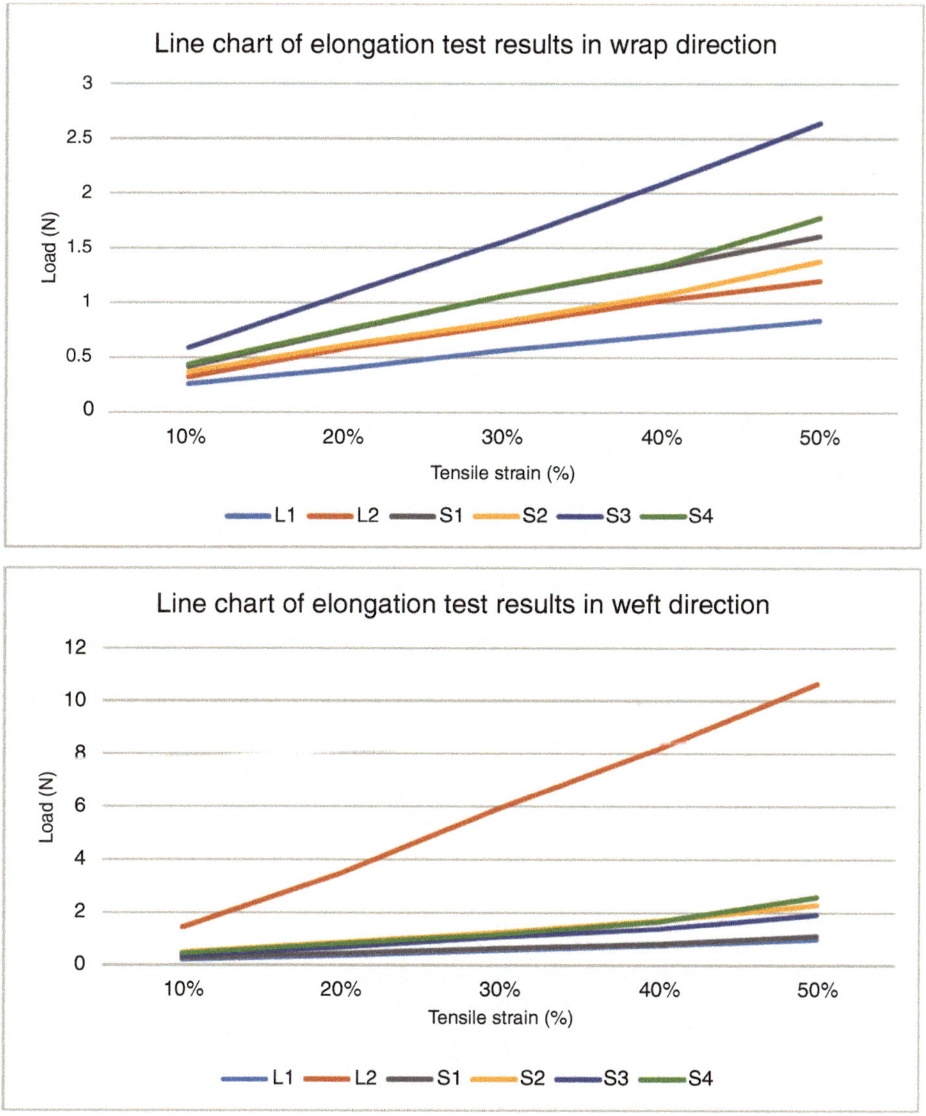

Figure 10.9 Line chart of elongation test results in warp and weft directions.

Resistance and Other Related Surface Changes of Textile Fabric Random Tumble Pilling Tester, to examine the resistance of the primary fabric used for the tank-top to the formation of pills and other related surface changes. Three samples of each fabric were tested, and the rate of pilling is shown in Fig. 10.11. A rating of 5 means that there is no pilling, and 1 is severe pilling.

In terms of the lining fabrics, L1 was found to have no pilling after the testing. All three L1 specimens scored a rating of 5 which means that the fabric resists the formation of pills when it randomly comes into contact with other surfaces.

Table 10.6 Air permeability test results

Fabric name	Air pressure (kPa.s/m)				
	Specimen	Specimen 2	Specimen 3	Mean	S.D.
LI	0.007	0.006	0.006	0.007	0.001
L2	0.026	0.027	0.0277	0.027	0.001
SI	0.634	0.806	0.623	0.688	0.084
S2	0.324	0.280	0.286	0.297	0.019
S3	0.723	0.845	0.831	0.800	0.055
S4	0.840	0.790	0.740	0.790	0.041

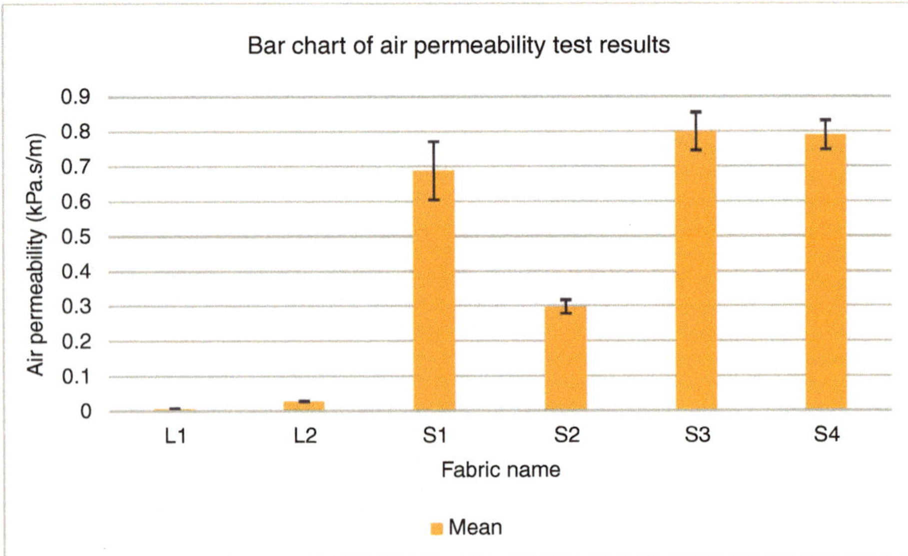

Figure 10.10 Bar chart of air permeability test results.

As for the shell fabrics, S1 has the lowest resistance to pilling in terms of random surface abrasion. This fabric has more severe pilling among all of the other fabrics. Therefore, S1 is not suitable for use as the primary fabric for the tank top because the garment is worn for a lengthy amount of time and therefore needs to be durable. S2 and S4 both provide a good result which means they have good resistance against random surface abrasion, with an accumulated rating of 13 for both fabric types. Therefore, with regard to the durability in terms of random tumble pilling resistance, both S2 and S4 are better options for the tank top as the appearance would not be influenced so much by pilling.

Another test that is important to determine the durability and fit of the tank-top is fabric dimensional stability after laundering which was conducted by using the AATCC Test Method 135- Dimensional Changes of Fabrics after Home Laundering. The test results are plotted in Fig. 10.12.

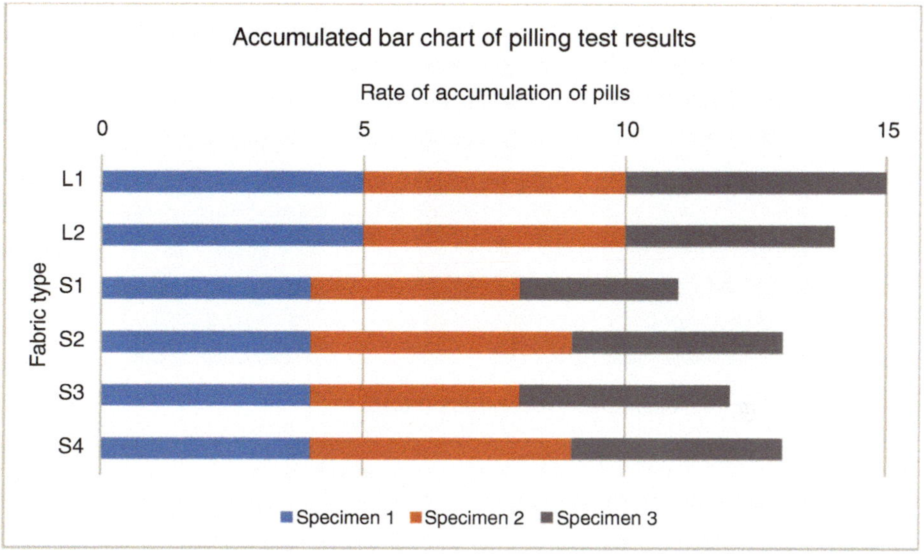

Figure 10.11 Pilling test results.

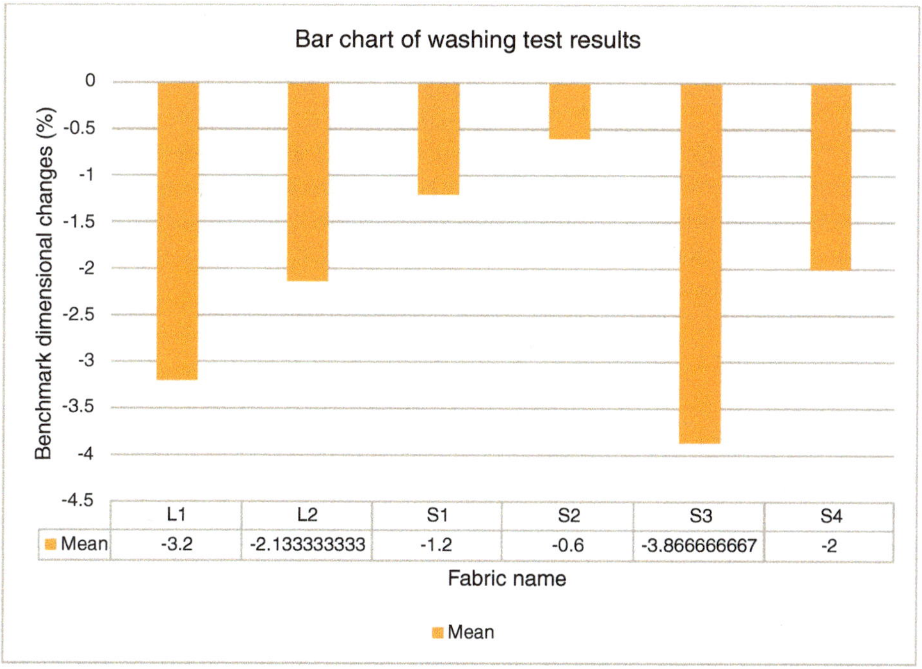

Figure 10.12 Washing test results.

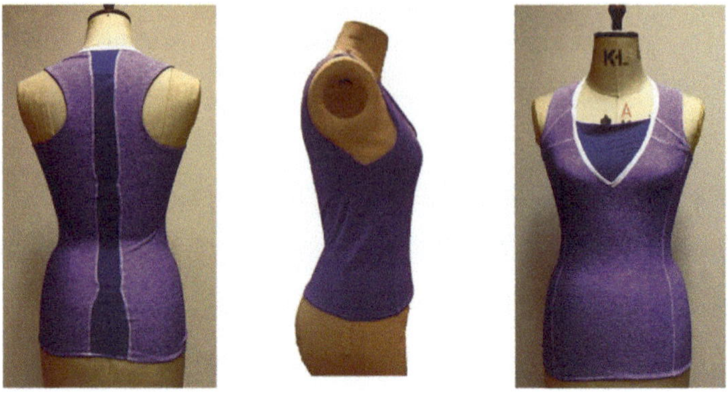

Figure 10.13 Tank top assembled with optimal fabrics.

The test results indicate that all of the fabrics have a small reduction in their dimensions after home laundering. In terms of the lining fabrics, L2 shrinks less than L1, with a shrinkage of 2.13% after it is laundered, and thus L2 shows a higher dimensional stability.

As for the shell fabrics, S1 and S2 show less shrinkage than the other fabrics. S2 has an average shrinkage of 0.6% after the testing, which means that it is a very good fabric for use as the shell of the tank-top as there is little shrinkage after washing.

Therefore, the pattern of the tank-top does not require adjustments if S2 is used due to its dimensional characteristics.

An overall evaluation based on the test results discussed above shows that L1 is more suitable for the tank top as a lining fabric because it is a four-way stretch material and can provide adequate elasticity in all directions to accommodate the different sizes of the foam cups, despite that the scores of both L1 and L2 are the same.

In terms of the shell fabrics, S2 has the best performance with relatively higher elongation, air permeability, pilling resistance, and dimensional stability after home laundering than the other fabrics. Considering the overall results, S2 is the most durable and comfortable, and therefore, was chosen for use as the shell fabric of the tank top.

In accordance with the physical textile laboratory test results, the tank top was assembled by using the optimal fabrics and is shown in Fig. 10.13.

10.5.5 Posture monitoring sensors

The design and material for the tank-top were determined based on body-mapping techniques and textile material testing. The tank-top itself does not have the ability to monitor the posture of the wearer; this task is left to the sensors. It is therefore important to incorporate posture monitoring sensors into the tank-top which would provide a complete system to monitor the posture of the wearers.

The sensors comprised a 3-axis accelerometer, gyroscope, thermometer and barometer. All of the functioning chips were mounted onto a micro controller unit (MCU) that was equipped with a Bluetooth connection system. The MCU is circular in shape

Figure 10.14 Micro controller unit.

with a diameter of 3 cm. An individual 3 V battery was embedded onto the back of the MCU (see Fig. 10.14). The battery allowed the sensor to function for at least 56 hours, which therefore, ensured that no battery changes were needed on a daily basis.

A mobile application was designed to provide visual information obtained from the posture monitoring sensors. The mobile application is compatible with most Android devices which feature Bluetooth 3.0. The application is able to connect and display the information of the 3 posture monitoring sensors at the same time. The application also shows the atmosphere temperature, body posture angle detected from each individual sensor, battery level and calibration of the sensors. The calibration function is a key element for synchronizing the data of the sensors with the sEMG training program. The sEMG training program developed provides visual sensory feedback triggered by sEMG signals. The posture monitoring sensors are synchronized through the sEMG signals which calibrate the sensors at that time. Therefore, the user is restricted to that position while her sEMG signal remains the same.

10.5.6 Wear trial and evaluation

The training session summary provided a means to examine the effectiveness of the sEMG posture training sessions. Customized sEMG posture training with the use of posture monitoring sensors serves two purposes for the users. First, the system allows the subjects to balance their muscle activity with less effort by reducing the difference between the sEMG signals on the left and right sides of the tested muscle regions. Second, the system reduces the sEMG signal value. In order to train the users to achieve the mentioned objectives, visual feedback was provided if the user fulfilled the requirements.

The sEMG signals processed from each training session with 12 participants were combined, and a comparison of the data before and after the sEMG posture training was carried out. The result of the posture training is shown in Table 10.7. The table presents the mean of the RMS sEMG value (mV) before and after the posture training session. For ease of comparison, a bar chart of the results is plotted and shown in Fig. 10.15.

Two observations can be made based on the result derived from the sEMG posture training data. The first observation is that the participants have a relatively lower

Table 10.7 RMS sEMG values before and after training

		Mean RMS sEMG value (S.D) (mV)			
N	Muscle region	Habitual sitting (Left)	Habitual sitting (Right)	Trained sitting (Left)	Trained sitting (Right)
12	Trapezius	5.365 (9.771)	9.619 (11.68)	3.336 (0.524)	2.974 (0.489)
	Latissimus Dorsi	7.465 (8.521)	5.84 (3.137)	3.569 (0.897)	3.881 (0.730)
	Erector Spinae Thoracic	7.066 (5.356)	11.78 (12.94)	2.808 (0.527)	3.176 (0.684)
	Erector Spinae Lumbar	5.157 (4.480)	15.09 (25.81)	2.776 (0.474)	3.093 (0.432)

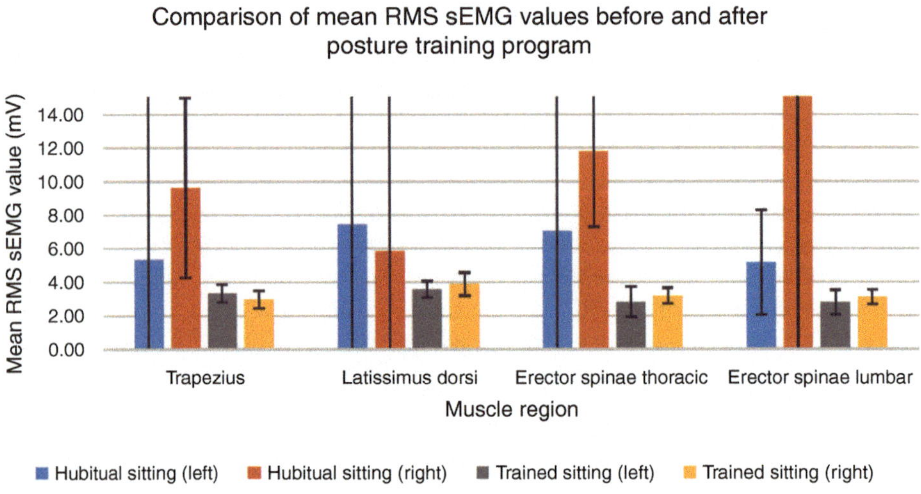

Figure 10.15 Comparison of mean RMS sEMG values before and after training.

sEMG activity while sitting after the training. Before the training, none of the mean RMS sEMG values are less than 5 mV among all of the tested muscle regions. The highest value is on the right side of the erector spinae at the lumbar region which recorded an average of 15.09 mV. This indicates that the participants have higher muscle activity when they are sitting. After around 30 sessions of posture training, the mean RMS sEMG value significantly declines with none of the tested muscle regions recording a value that is higher than 4 mV. The highest value of 3.881 mV is on the right side of the latissimus dorsi. This result fulfills one of the objectives of the sEMG posture training, which is to train the participants to complete the same motion of sitting with reduced muscle activity. Kippers and Parker (Kippers & Parker, 1985) suggested that fatigue is caused by constant static muscular work. It is recommended that during sitting, the level of muscle activity should be minimized at all possible levels. The result indicates that biofeedback training can help the user to relax the target muscle regions through self-regulation. The second observation is that the standard

Table 10.8 Mean RMS sEMG ratio for sitting

		Mean RMS sEMG sitting ratio (S.D.)	
N	**Muscle regions**	**Before training**	**After training**
8	Trapezius	2.30 (1.75)	0.95 (0.13)
	Latissimus Dorsi	1.14 (0.50)	1.12 (0.24)
	Erector Spinae Thoracic	2.17 (2.29)	0.97 (0.20)
	Erector Spinae Lumbar	1.00 (0.44)	0.97 (0.13)

deviation is lower after the training. This means that the data distribution across the 12 subjects is comparatively more evenly distributed than before the training. The training successfully trained the users to maintain a certain level of muscle activity throughout the training sessions. It should be noted that the mean RMS sEMG value is not meant to be used for direct comparison among the different individuals as the data were not normalized, and the maximum voluntary contraction of each muscle region of each subject may differ.

Therefore, the sEMG value is calculated as an sEMG ratio to normalize the data for analysis.

In Table 10.8, the result of the mean RMS sEMG ratio is shown and a bar chart for ease of comparison is plotted in Fig. 10.16. The sEMG training also serves another purpose, which is to train the users to sit in a more balanced posture as denoted by an sEMG signal. The mean RMS sEMG ratio (Chwała et al., 2014) is one of the determinants to indicate the level of symmetrical muscle activity during sitting. The users should ideally obtain a ratio of 1 in which the left and right sides of the same muscle region have perfectly symmetrical muscle activity. From the data obtained before and after the training program, the mean ratios at all tested muscle regions except for the erector spinae at the lumbar regions showed an improvement in terms of the sEMG ratio for sitting after 30 sessions of training. Therefore, the subjects have a relatively more balanced posture after the training. A statistical analysis was performed to examine the significance of the data, and a repeated measures analysis of variance (ANOVA) with the factor of time (before and after training) and four muscle regions was conducted. This was endeavored to determine whether the ratio before and after the posture training is statistically significantly different. From the repeated measures ANOVA, a trend of significant difference before and after posture training was found, with a Wilks' lambda = 0.574, $F(1, 7) = 5.20$, and $P = 0.057$. This indicates that the training has a time effect when comparing the data before and after the posture training. Numerically, the subjects are able to achieve a relatively more balanced sitting posture as demonstrated by the sEMG signals. The purpose of the training is therefore realized. Even though the suggested therapy for mild idiopathic scoliosis is carrying out periodical observations rather than the prescription of any type of treatment or exercise during the early stages unless a rapid change takes place in the curvature angle or there is spine rotation, however, there is the possibility of treatment with the use of exercise, as indicated in the literature (Lau, 2011; Mehta, 1978; Monroe, 2012). In 2003, Hawes (Hawes, 2003) provided a critical review on the use of exercise for the

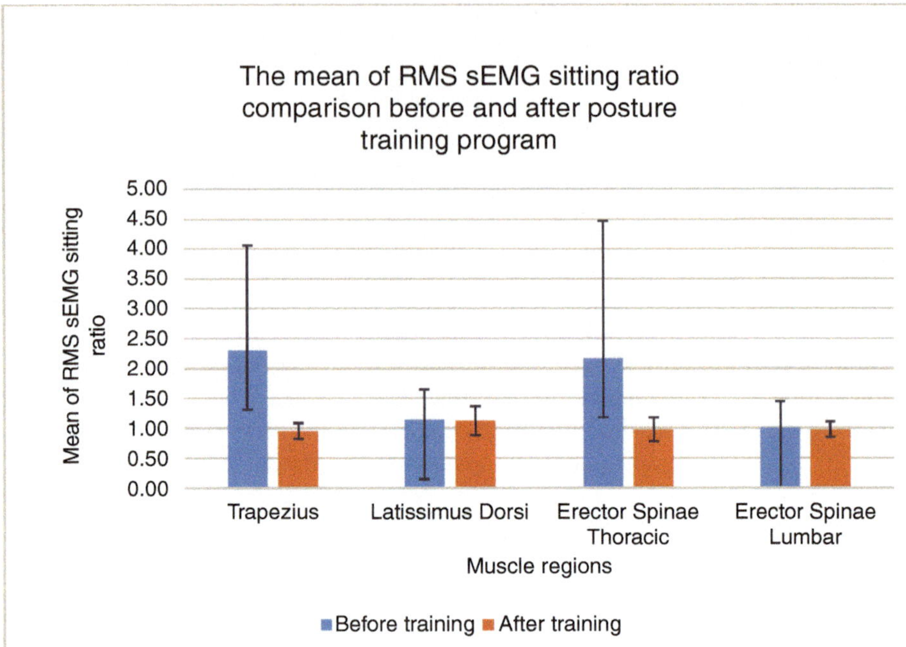

Figure 10.16 Comparison of mean RMS sEMG ratio for sitting before and after training.

treatment of scoliosis and examples of successful cases in which scoliosis was treated with exercise, such as work by Dickson et al., (1984). By adding loads to recover the postural balance of the patients: "the spinal deformity can be completely eliminated" (Hawes, 2003).

Chwała et al. (2014) also performed an sEMG assessment for adolescents with idiopathic scoliosis and suggested that asymmetric load-free and symmetric exercises are beneficial in corrective endeavors. This study provides a similar finding in that behavioral training is effective for training muscle activity toward an optimum level of activity. Through motor learning, those with AIS can permanently maintain postural balance by practicing the recommended positions on a regular basis as a new motor task (Dickson et al., 1984).

10.6 Conclusion

The problems that have led to the implementation of this preliminary work are the lack of options in passive treatment for mild scoliosis patients other than regular checkups, and therapy with either rigid or flexible braces which affects the quality of life of scoliosis patients. There are also few options in biofeedback training that have been specifically developed for AIS patients. However, the use of a biofeedback system which can monitor and alert scoliosis patients means that they can have better control over their own body, including maintaining the correct body posture and controlling their spinal deformity.

Improving the quality of life of a patient while undergoing bracing treatment should be considered a priority. This study therefore combines clinical experience with textile and materials sciences, and provides a monitoring process that uses biofeedback electromyography (sEMG) to train the muscles. Based on such, a posture training tank-top for adolescents with early scoliosis has been developed. The resultant posture training tank-top has significant potential, and not only the ability to provide posture monitoring through sEMG muscle training for adolescents on a regular basis, but also accommodates sensors that can monitor the posture coordinates on a real time basis and track the progress of the posture training. As a result, the future likelihood of brace wear or surgery will be greatly reduced.

Background information on scoliosis has been provided here with a comprehensive literature review. Taking into account the statistical results, AIS in females should be the focus as they are the most at risk. In addition, a knowledge gap is found in that the treatment options for mild scoliotic patients (spinal curvature of 10–20°) are passive. Only periodic observation is recommended. In terms of the those with spinal curvature of 21–45°, the effectiveness of bracing treatment is fair and very much depends on the compliance of the users, as the quality of life of the patients may be affected. Therefore, treatment that focuses on the well-being of patients should be given priority. In order to determine how a reliable sEMG training program could be developed, which would also take into consideration the well-being of the patients, a review on the application of sEMG has been performed. It is found that two studies have used biofeedback postural training with scoliotic subjects. The strengths and weaknesses of the two studies are compared and taken into consideration for the posture training tank-top in this study.

The methods used for this study have also been discussed. The entire experimental design comprises two phases, namely: (1) the design and development of a posture training tank-top and (2) a wear trial and evaluation of the posture training tank-top. In the first phase of the study, a posture training tank-top design that can accommodate posture monitoring sensors is developed based on the design requirements. The tank top is tight-fitting with two shoulder straps designed to minimize unwanted movement that might affect the reliability of the posture monitoring sensors. The materials for the tank-top are also selected based on various fabric test results. For the shell fabric of the tank-top, a tricot fabric is selected as it shows relatively better test results than the other fabrics. This fabric is able to maximize comfort, durability and fit among other factors. A powernet fabric is selected as the lining fabric. Although this powernet fabric produces similar test results as the satinnet fabric, its four-way stretch property can accommodate different foam cup sizes. Therefore, a powernet fabric (L1) is selected. Three sensors are placed at T3, T12 and the L4-L5 joints to monitor the posture of the user. The reliability of the posture sensors is examined with various initial tests. The result indicates that different detectors within the sensor are able to interact well with each other and provide reliable functioning. Even though the sensors at times lost connection with the

Android device, they are still reliable for training purposes as the sensors are able to automatically reconnect with the smart phone.

Twelve subjects have participated the posture training program which comprises the second phase of the study. After 30 sessions of training, the subjects are able to retain a relatively more balanced sEMG signal along their paraspinal muscles. It is found that, before the training, the sEMG ratio at the trapezius region shows a statistically significant difference ($P = 0.038$) with a test value of 1 and after the training, none of the tested regions show this trend. Through motor learning, those with AIS can permanently maintain postural balance by practicing the recommended positions on a regular basis as a new motor task.

References

Agabegi, S. S., & Agabegi, E. D. (2005). *Step-Up to Medicine (Step-Up Series)*. Hagerstwon: Lippincott Williams & Wilkins.

Ahmed, M. U., Begum, S., Funk, P., Xiong, N., & von Scheele, B. (2011). A multi-module case-based biofeedback system for stress treatment. *Artificial Intelligence in Medicine*, *51*(2), 107–115.

Avikainen, V. J., Rezasoltani, A., & Kauhanen, H. A. (1999). Asymmetry of paraspinal EMG-time characteristics in idiopathic scoliosis. *Journal of Spinal Disorders & Techniques*, *12*(1), 61–67.

Bazzarelli, M., Durdle, N., Lou, E., & Raso, J. (2001). A low power portable electromagnetic posture monitoring system. *Proceedings of the 18th IEEE Instrumentation and Measurement Technology Conference*, 1, pp. 619–623.

Berger, N., Edelsten, J., Fishman, S., Krebs, D., & Springer, W. (1983). *Spinal orthotics: Prosthetics and orthotics*. New York: New York University, Post-Graduate Medical School.

Bray, D. (1998). Biofeedback. *Complementary Therapies in Nursing and Midwifery* (1), 22–24.

Bunge, E. M., de Bekker-grob, E. W., van Biezen, F. C., Essink-bot, M. L., & de Konig, H. J. (2009). Patients' preferences for scoliosis brace treatment. A discrete choice experiment. *Spine*, *35*, 57–63.

Cassella, M., & Hall, J. E. (1991). Current treatment approaches in the nonoperative management of adolescent idiopathic scoliosis. *Physical Therapy*, *71*, 897–909.

Chwała, W., Koziana, A., Kasperczyk, T., Walaszek, R., & Płaszewski, M. (2014). Electromyographic assessment of functional symmetry of paraspinal muscles during static exercises in adolescents with idiopathic scoliosis. *BioMed Research International*.

Coillard, C., Circo, A. B., & Rivard, C. H. (2010). SpineCor treatment for Juvenile Idiopathic Scoliosis: SOSORT award 2010 winner. *Scoliosis Journal*, *5*(25), 1–7.

Collier, B., & Epps, H. (1999). *Textile testing and analysis*. Upper Saddle River, N.J: Merrill.

Deceuninck, J., & Bernard, J. C. (2012). Quality of life and brace-treated idiopathic scoliosis: a cross-sectional study performed at the Centre des Massues on a population of 120 children and adolescents. *Annals of Physical and Rehabilitation Medicine*, *55*(2), 93–102.

Dickson, R. A., & Bradford, D. S. (1984). In *Management of spinal deformities* (Vol. 2). Butterworth-Heinemann.

DiRaimondo, C. V., & Green, N. E. (1988). Brace-wear compliance in patients with adolescent idiopathic scoliosis. *Journal of Pediatric Orthopaedics*, *8*, 143–146.

Dobbs, M. B., & Weinstein, S. L. (1999). Infantile and juvenile scoliosis. *Orthopedic Clinics of North America, 30*(3), 331–341.

Dolan, L. A., Donnelly, M. J., Spratt, K. F., & Weinstein, S. L. (2007). Professional opinion concerning the effectiveness of bracing relative to observation in adolescent idiopathic scoliosis. *Journal of Pediatric Orthopaedics, 27*(3), 270276.

Dworkin, B., Miller, N., Dworkin, S., Birbaumer, N., Brines, M. L., Jonas, S., Schwentker, E. P., & Graham, J. J. (1985). Behavioral method for the treatment of idiopathic scoliosis. *Medical Sciences, 82*(8), 2493–2497.

Fallstrom, K., Nachemson, A. L., & Cochran, T. P. (1984). Psychologic effects of treatment for adolescent idiopathic scoliosis. *Orthop Trans, 8*, 150.

Gatchel, R. J., & Price, K. P. (1979). *Clinical applications of biofeedback: Appraisal and status*. New York: Pergamon Press.

Hasler, C. C., Wietlisbach, S., & Büchler, P. (2010). Objective compliance of adolescent girls with idiopathic scoliosis in a dynamic SpineCor brace. *Journal of Children's Orthopaedics, 4*, 211–218.

Hawes, M. C. (2003). The use of exercises in the treatment of scoliosis: An evidence based critical review of the literature. *Developmental Neurorehabilitation, 6*(3–4), 171–182.

Helfenstein, A., Lankes, M., Ohlert, K., Varoga, D., Hahne, H. J., Ulrich, H. W., & Hassenpflug, J. (2006). The objective determination of compliance in treatment of adolescent idiopathic scoliosis with spinal orthoses. *Spine (Phila Pa 1976), 31*, 339–344.

Hewitt, J. (2011). *Scoliosis: Causes, tests and treatments*. Charleston, SC: CreateSpace.

Hong, Y., Fong, D. T. P., & Li, J. X. (2011). The effect of school bag design and load on spinal posture during stair use by children. *Ergonomics, 54*(12), 1207–1213.

James, J. I. (1954). Idiopathic scoliosis: the prognosis, the diagnosis and operative indications related to curve patterns and the age of onset. *The Journal of Bone and Joint Surgery British Volumc, 36*, 35–49.

Juskeliene, V., Magnus, P., Bakketeig, L. S., Dailidiene, N., & Jurkuvenas, V. (1996). Prevalence and risk factors for asymmetric posture in preschool children aged 6-7 years. *International Journal of Epidemiology, 25*(5), 1053–1059.

Kane, W. J. (1997). Scoliosis prevalence: A call for a statement of terms. *Clin Orthop., 126*, 43–46.

Kappes, B. (2008). Biofeedback therapy: Training or treatment. *Applied Psychophysiology and Biofeedback, 33*, 173–179.

Kippers, V., & Parker, W. (1985). Electromyographic studies of erectores spinae: Symmetrical postures and sagittal trunk motion. *Australian Journal of Physiotherapy, 31*(3), 95–105.

Korovessis, P., Zacharatos, S., Koureas, G., & Megas, P. (2007). Comparative multifactorial analysis of the effects of idiopathic adolescent scoliosis and *Scheuermann kyphosis* on the self-perceived health status of adolescents treated with brace. *European Spine Journal, 16*, 537–546.

Lau, K. (2011). *Health in your hands: your plan for natural scoliosis prevention and treatment* (2nd ed.). Singapore: Health in Your Hands Pte.

Lehrer, P., Smetankin, A., & Potapova, T. (2000). Respiratory Sinus Arrhythmia Biofeedback Therapy for Asthma: A Report of 20 Unmedicated Pediatric Cases Using the Smetankin Method. *Applied Psychophysiology and Biofeedback*(3), 193–200.

Lindeman, M., & Behm, K. (1999). Cognitive strategies and self-esteem as predictors of brace-wear noncompliance in patients with idiopathic scoliosis and kyphosis. *Journal of Pediatric Orthopaedics, 19*, 493–499.

Lonstein, J. E., & Carlson, J. M. (1984). The prediction of curve progression in untreated idiopathic scoliosis during growth. *Journal of Bone and Joint Surgery [America], 66*, 1061.

Lou, E., Raso, J. V., Hill, D. L., Mahood, J. K., & Moreau, M. J. (2004). Cor- relation between quantity and quality of orthosis wear and treatment outcomes in adolescent idiopathic scoliosis. *Prosthetics and Orthotics International*, *28*, 49–54.

Lyons, G.M., Sharma, P., Baker, M., O'Malley, S., & Shanahan, A. (2003, September). A computer game-based EMG biofeedback system for muscle rehabilitation. In Engineering in Medicine and Biology Society, 2003. Proceedings of the 25th Annual International Conference of the IEEE 2, 1625-1628). IEEE.

Mackenzie, F. A. (1922). The operative treatment of scoliosis. *The Journal of Bone & Joint Surgery*, *4*(3), 446–452.

Marschall, M., Harrington, A. C., & Steele, J. R. (1995). Effect of work station design on sitting posture in young children. *Ergonomics*, *38*(9), 1932–1940.

Martin, E. (Ed.). (1998). *Oxford concise colour medical dictionary* (2nd ed.). Oxford, New York: Oxford University Press.

Matthews, M., & Crawford, R. (2006). The use of dynamic *Lycra orthosis* in the treatment of scoliosis: A case study. *Prosthetics and Orthotics International*, *30*(2), 174–181.

McKenzie, R. (1980). *Treat Your Own Back.* Spinal Publications New Zealand Ltd.

Mehta, M. (1978). Pain provoked scoliosis. Observations on the evolution of the deformity. *Clinical Orthopaedics and Related Research*(135), 58.

Monroe, M. (2012). *Yoga and scoliosis: A journey to health and healing.* New York, NY: Demos Health.

Morgan, T. H., & Scott, J. C. (1956). Treatment of infantile idiopathic scoliosis. *The Journal of Bone and Joint Surgery. British volume*, *38*, 450–457.

Morton, A., Riddle, R., Buchanan, R., Katz, D., & Birch, J. (2008). Accuracy in the prediction and estimation of adherence to bracewear before and during treatment of adolescent idiopathic scoliosis. *Journal of Pediatric Orthopaedics*, *28*, 336–341.

Myers, B. A., Friedman, S. B., & Weiner, I. B. (1970). Coping with a chronic, disability: Psychosocial observations of girl with scoliosis treated with a Milwaukee brace. *The American Journal of Diseases of Children*, *120*, 175–181.

Nestoriuc, Y., Martin, A., Rief, W., & Andrasik, F. (2008). Biofeedback treatment for headache disorders: A comprehensive efficacy review. *Applied Psychophysiology and Biofeedback*, *33*(3), 125–140.

Nicholson, G. P., Ferguson-pell, M. W., Smith, K., Edgar, M., & Morley, T. (2002). Quantitative measurement of spinal brace use and compliance in the treatment of adolescent idiopathic scoliosis. *Studies in Health Technology and Informatics*, *91*, 372–377.

Nicholson, G. P., Ferguson-pell, M. W., Smith, K., Edgar, M., & Morley, T. (2003). The objective measurement of spinal orthosis use for the treatment of adolescent idiopathic scoliosis. *Spine (Phila Pa 1976)*, *28*, 2243–2250.

Oates, S., Evans, G. W., & Hedge, A. (1998). An anthropometric and postural risk assessment of children's school computer work environments. *Computers in the Schools*, *14*(3–4), 55–63.

Ólafasson, Y., Saraste, H., & Ahlgren, R. M. (1999). Does bracing affect self-image? A prospective study on 54 patients with adolescent idiopathic scoliosis. *European Spine Journal*, *12*, 141–148.

Pehrsson, K., Larsson, S., Oden, A., & Nachemson, A. (1992). Long-term follow-up of patients with untreated scoliosis A study of mortality, causes of death, and symptoms. *Spine*, *17*(9), 1091–1096.

Prosthetic and Orthotic Unit, Hong Kong Sanatorium & Hospital, 2005, Retrieved from: http://www.hksh.com/pdf/pamphlet/po/P_OF2A_Nov08.pdf.

Reamy, B. V., & Slakey, J. B. (2001). Adolescent Idiopathic Scoliosis: Review and current concepts. *American Family Physician*, *64*(1), 111–116.

Roach, J. W. (1999). Adolescent idiopathic scoliosis. *Orthopedic Clinics of North America, 30*(3), 353–365.

Schleenbaker, R. E., & Mainous, A. G., III (1993). Electromyographic biofeedback for neuromuscular reeducation in the hemiplegic stroke patient: A metaanalysis. *Medicine, 40536*, 0284.

Smith, C., & Havenith, G. (2011). Body mapping of sweating patterns in male athletes in mild exercise-induced hyperthermia.

Smith, C., & Havenith, G. (2012). Body mapping of sweating patterns in athletes: A sex comparison.

Tones, M., Moss, N., & Polly, D. W. (2006). A review of quality of life and psychosocial issues in scoliosis. *Spine, 31*, 3027–3038.

USC Center for Spinal Surgery. (2005). Scoliosis, Children's spine surgery. USC Center for Spinal Surgery. Accessed 20 August 2012. http://www.uscspine.com/conditions/childrens-scoliosis.cfm.

Webster's New Universal Unabridged Dictionary: Deluxe. (1983). New York: Simon & Schuster.

Weiss, H. R., & Weiss, G. M. (2005). Brace treatment during pubertal growth spurt in girls with idiopathic scoliosis (IS): A prospective trial comparing two different concepts. *Pediatric Rehabilitation, 8*(3), 199–206.

Wickers, F. C., Bunch, W. H., & Barnett, P. M. (1977). Psychological factors in failure to wear the Milwaukee brace for treatment of idiopathic scoliosis. *Clinical Orthopaedics, 126*, 62–66.

Wong, W. (2009). Development of a posture monitoring system. PhD thesis. Hong Kong: Dept. of Health Technology and Informatics, The Hong Kong Polytechnic University. Available from: http://ira.lib.polyu.edu.hk/handle/10397/2774.

Wong, W. Y., & Wong, M. S. (2008). Smart garment for trunk posture monitoring: A preliminary study. *Scoliosis, 3*(7), 1–9.

Wong, M. S., Mak, A. F. T., Luk, K. D. K., Evans, J. H., & Brown, B. (2001). Effectiveness of audio-biofeedback in postural training for adolescent idiopathic scoliosis patients. *Prosthetics and Orthotics International, 25*(1), 60–70.

Wong, M. S., Cheng, C. Y., Lam, T. P., Ng, K. W., Sin, S. W., Lee-Shum, L. F., Chow, H. K., & Tam, Y. P. (2008). The effect of rigid versus flexible spinal orthosis on the clinical efficacy and acceptance of the patients with adolescent idiopathic scoliosis. *Spine, 33*(12), 1360–1365.

Zacharkow, D. (1988). *Posture : sitting, standing, chair design and exercise*. Thomas.

Index

Note: Page numbers followed by "f" indicate figures, "t" indicate tables.

FSC
www.fsc.org
MIX
Papier | Fördert gute Waldnutzung
FSC® C083411

Kolibri 360 GmbH
Ferdinand-Jühlke-Straße 7
99095 Erfurt, Deutschland
produktsicherheit@kolibri360.de